T0330089

Regulating Offshore Petroleum Resources

Regulating Offshore Petroleum Resources

The British and Norwegian Models

Edited by

Eduardo G. Pereira

University of Eastern Finland, Finland and Scandinavian Institute of Maritme Law, University of Oslo, Norway

Henrik Bjørnebye

Scandinavian Institute of Maritime Law, University of Oslo, Norway

Cheltenham, UK • Northampton, MA, USA

Published by
Edward Elgar Publishing Limited
The Lypiatts
15 Lansdown Road
Cheltenham
Glos GL50 2JA
UK

Edward Elgar Publishing, Inc.
William Pratt House
9 Dewey Court
Northampton
Massachusetts 01060
USA

A catalogue record for this book
is available from the British Library

Library of Congress Control Number: 2019945329

This book is available electronically in the **Elgar**online
Law subject collection
DOI 10.4337/9781785368912

ISBN 978 1 78536 890 5 (cased)
ISBN 978 1 78536 891 2 (eBook)

Printed and bound by CPI Group (UK) Ltd, Croydon, CR0 4YY

Contents

Contributors

Catherine Banet – Associate Professor at the Scandinavian Institute of Maritime Law, University of Oslo, Norway

Henrik Bjørnebye – Professor at the Scandinavian Institute of Maritime Law, University of Oslo, Norway

Yanal Abul Failat – Associate Lawyer, LXL LLP, UK

Greg Gordon – Senior Lecturer at the School of Law, University of Aberdeen, UK

Tonje Pareli Gormley – Partner in the Oil, Offshore and Energy Department, Arntzen de Besche, Norway

Mohammad Hazrati – Research Assistant, University of Dundee, UK

Raphael Heffron – Professor in Global Energy Law and Sustainability at the Centre for Energy, Petroleum and Mineral Law and Policy, University of Dundee, UK

Erik Jarlsby – Partner, Eureka Energy Partners AS, Norway

Merete Kristensen – Senior Associate in the Oil, Offshore and Energy Department, Arntzen de Besche, Norway

Darren McCauley – Senior Lecturer, University of St Andrews, UK

Eduardo G. Pereira – Professor of Natural Resources and Energy Law, Externado University of Colombia, Colombia and Researcher, University of Oslo, Norway

PART I

Introduction

1. Introduction

Eduardo G. Pereira and Henrik Bjørnebye

Section 1–2 of the Norwegian Petroleum Act sets out that petroleum resource management:

> shall be carried out in a long-term perspective for the benefit of the Norwegian society as a whole. In this regard the resource management shall provide revenues to the country and shall contribute to ensuring welfare, employment and an improved environment, as well as to the strengthening of Norwegian trade and industry and industrial development, and at the same time take due regard to regional and local policy considerations and other activities.[1]

Other states use different wording to describe their aims, but in substance most would agree that the description above captures the overall purpose of petroleum resource management. The greater challenge is how to achieve these aims. Different models are applied from country to country, depending on the societal, political, economic and legal context.

At the same time, the global petroleum industry has paved the way for standardization across jurisdictions. Petroleum states – in particular, those in emerging markets – may wish to adopt international best practice in order to attract investment. But what is international best practice for petroleum resource management? And to what extent can solutions successfully implemented in one country be replicated by another country with other societal conditions?

In this book we intend to contribute to the understanding of these questions by analysing the regulatory models of two mature provinces with longstanding petroleum resource management experience: Norway and the UK. Through this analysis, we seek to explain why and how the Norwegian and British governments decided to develop their natural resources in the manner chosen. The lessons learned during this process are of interest to other countries for several reasons. The management models of both countries are based on nearly 50 years' experience and are considered robust internationally. Moreover, although the two countries manage resources that share the same border, they

---

[1]   Petroleum Act 29 November 1996 No 72, Section 1–2, second paragraph. Translation by the Norwegian Petroleum Directorate, available at www.npd.no.

have chosen distinctly different ways to govern their sectors. Finally, in an era when rapid energy transition is needed to avoid climate change, the extensive experience gained by Norway and the UK in developing petroleum law and policy may also provide valuable lessons for the development of future non-fossil energy resource management regimes.

The book consists of a Norwegian part and a UK part, each divided into three chapters.

The first chapter of each part describes the societal context in which each jurisdiction's petroleum resource management systems were developed. This includes a broad description of the economy and industries, human resources, geography, access to the market, demographics, infrastructure, welfare systems, culture (including anti-corruption and transparency practices), cooperation with the industry, politics and environmental standards. Any resource management regime must take due account of these aspects in order to succeed. Consequently, inspiration and lessons learned drawn from one system must always be considered on the basis of differences in societal contexts. For this reason, we have chosen to include detailed chapters describing the state of development in Norway and the UK at the time their offshore petroleum resources were discovered and later developed.

The second chapter in each part is devoted to the development of petroleum policy and law in Norway and the UK, respectively.

Norwegian petroleum policy and law have been characterized by strong host country control based on a progressively developed licence system and significant state ownership. The second Norwegian chapter analyses this regime, commonly referred to as the 'Norwegian model', showing how it has been implemented and further developed within the societal context described in the first country chapter. The authors argue that the manner in which the model has been implemented and developed over the past decades to fit the specific needs of the sector and broader society has proven to be its strength.

The second UK chapter describes the evolution of British petroleum policy and law, including state governance and government take, as well as public facilitation of investments on the UK Continental Shelf. The authors provide a critical account of how several factors – such as lack of strategic planning and competing political ideologies – have until recently resulted in a system of non-interventionist governance, which has arguably had several negative effects for the development of the UK petroleum sector.

Based on the broader petroleum law and policy perspectives provided in the second chapters, the third and final chapter in each part considers in more detail the legal structure of the national resource management models. These chapters focus on the content of the licence systems and their relationship to joint operating agreements as important resource management tools. The chapters also provide an overview of the sector regulation of environment and climate commitments.

Finally, in the concluding part of the book, we seek to draw some overall conclusions from the comparison of the Norwegian and UK models.

PART II

Norway

2.   Background: Norway

Erik Jarlsby

2.1   INTRODUCTION

The aim of this chapter is to describe the societal context in which Norway's system for managing its petroleum resources was developed. Norwegian society is briefly described in terms of economy, resources and policies.

The first significant policy decisions concerning possible Norwegian petroleum resources were made in 1963 and the first discovery was made in 1968. This chapter therefore focuses on the period from the 1960s to the present. Of particular interest for the purposes of this book are the 1960s and 1970s, during which the foundations of Norwegian petroleum policy were laid.

Section 2.2 reviews Norway's economy and industries, with an emphasis on those sectors which were clearly helpful in facilitating the development of the petroleum sector. Section 2.3 examines Norway's human resources and skill sets, from which the petroleum sector could draw capable personnel; it is followed by a fourth section on demographics. Infrastructure and market access for businesses are reviewed in section 2.5. The chapter then turns to the less tangible aspects of the context in which Norway's petroleum sector and its management have evolved. Sections 2.6–9 address the country's welfare system, cultural practices and transparency, political context and environmental standards. Section 10 reviews the relationships between industry stakeholders in Norway – notably between enterprises, the State and labour unions – which are reflected in developments in the petroleum sector. A final section concludes.

The statistical data referred in this chapter are sourced from Statistics Norway,[1] except where specified otherwise. Work on this chapter was completed on 10 January 2018, and the most recent statistics available at that time have been applied where relevant. The chapter also makes references to

---

[1]   www.ssb.no, last accessed 10 January 2018.

Norwegian laws which can be found on a common website,[2] albeit in most cases in the Norwegian language only.

2.2 ECONOMY AND INDUSTRIES

2.2.1 Overall Economic Position and Growth

In 1970, just before Norwegian oil production started, Norway's gross domestic product (GDP) per capita was USD 21 344 when measured at 2010 purchasing power parity.[3] This was 8 per cent less than the United States, similar to neighbouring Sweden and Denmark, and 20–29 per cent higher than France and UK, respectively, at the time. In the same GDP per capita terms, Norway as of 1970 was 8 per cent lower than Russia and Turkey as of 2016, but more advanced than Argentina, China and South Africa as of 2016 (by 18 per cent, 61 per cent[4] and 77 per cent, respectively). Norway thus already had a relatively advanced economy when it commenced petroleum production.

Norway's favourable economic position as of 1970 followed a period of strong and uninterrupted growth since the end of the Second World War. The country had been held back by German wartime occupation, devastation of the northernmost region and further destruction from the war elsewhere in the country. That damage was effectively repaired by the early 1950s. From the late 1940s to around 1970, the country implemented an economic policy targeted at promoting economic growth and increasing income through industrial investment, enhanced productivity and heightened competitiveness in export markets. Employment was much reduced in such traditional sectors as small-scale farming, fisheries, forestry (wood logging) and home-oriented industries. Employment grew in export-oriented industries, services and the public sector.

The period of uninterrupted post-war growth gave way to slower growth and occasional business downturns from the 1970s, in Norway as in other Western economies. International business cycles affected Norway quite significantly, as it is a small country with significant foreign trade and several export sectors exposed to international market fluctuations.

Nevertheless, from 1970–2016, Norway's economy grew by 180 per cent in terms of GDP per capita, still at purchasing power parity as per Organisation for Economic Co-operation and Development (OECD) data. It enjoyed

---

[2]   www.lovdata.no, last accessed 10 January 2018.

[3]   Organisation for Economic Co-operation and Development statistics, http://stats .oecd.org, last accessed 10 January 2018.

[4]   The comparison with China is for 2015, as the Chinese data for 2016 was not available at the time of writing.

stronger growth over that period than any larger Western European economy. By the same measure, Norway as of 2016 was 69 per cent more affluent than the European Union on average. The petroleum sector undoubtedly contributed much to this strong growth – both directly as an object of economic activity and indirectly by allowing for more expansive government finances than elsewhere in Europe.

As Norway is home to just 5.3 million people (2017), those parts of its economy which are exposed to international competition feature greater specialization in a few strong sectors than is usually found in larger industrialized economies. Some of these sectors, which were already at advanced stages by 1970, have developed competencies which have become valuable enablers for the development of the petroleum sector in Norway. The most important among these are reviewed below, followed by a review of the petroleum sector itself.

2.2.2   Shipping and Shipbuilding

Building and using ships has been an essential feature of Norwegian livelihood for more than 1000 years. Norway has had a large merchant shipping fleet since the nineteenth century, which has ranked as the world's third largest at times. There is still a large Norwegian shipping sector, even now that much of the fleet under Norwegian management operates under foreign flags of convenience. Overseas shipping contributed 9 per cent to Norway's gross national product (GNP) in 1970 and 0.8 per cent in 2017. The shipbuilding industry contributed 1.5 per cent to GNP in 1970. This industry has since expanded its scope to include construction of offshore petroleum installations and contributed 0.5 per cent to GNP in 2017.[5]

From the 1950s onwards, productivity in international shipping increased considerably due to larger ship sizes, better equipment and improved port operations. Norwegian ship owners contributed to this development, particularly in the oil tanker segment. By the mid-1970s, significant overcapacity had developed in this and other shipping segments, aggravated by the sharp increase in oil prices in 1974 and the reopening of the Suez Canal in 1975. Depressed shipping rates caused many ships to be laid up.

The downturn in the international shipping business caused orders for ship construction to be much diminished, affecting the yard industry. During the same period, however, the discovery of large oil and gas fields in the North Sea

---

[5]   The GNP related figures for 2017 referenced here are preliminary, based on four quarters ending with Q3 2017.

presented new opportunities for the yards. Large installations for the Ekofisk, Statfjord and other fields were built in Norway in the 1970s.

Those early Norwegian oil fields were at the global frontier in terms of developing petroleum resources in difficult offshore conditions. In terms of competencies, workers and firms from Norway's maritime sector made significant contributions to mastering these challenges and were well placed to enter the new petroleum sector with relative ease. Many merchant seamen, having lost employment in overseas shipping, found new employment on the offshore petroleum installations, including fixed platforms, floating rigs and the vessels supplying them. Shipbuilding skills were a useful basis for developing offshore petroleum installations.

Since the beginning of Norwegian petroleum operations, there have been significant overlaps between the shipping and petroleum industries. Beyond conventional transport of oil in tanker ships, there are tanker ships designed for loading crude oil directly from offshore installations. Floating rigs for drilling wells are in many cases owned and operated by firms with a background in shipping; as are ship-shaped installations for oil and gas production (floating production, storage and offtake units). A large fleet of vessels brings supplies for offshore petroleum operations and a variety of vessels are designed to perform special operations related to offshore petroleum production. Norwegian shipping firms are large providers of such ships. The markets are essentially global, as capabilities developed serving the Norwegian petroleum sector can readily be redeployed for offshore operations all over the world.

The industry for building ships and offshore installations largely upheld its real output value from 1970–2017, even if its relative contribution to GNP has diminished. It no longer builds oil tankers and other standard vessels, focusing instead on specialized, higher-value orders. In the 1970s–80s, Norwegian constructors of offshore installations gained decisive competitive advantage from their access to shielded deep waters in fjords. Technologies for offshore installations later evolved so as to make this particular advantage less relevant. Norwegian yards have no exclusive right to building Norwegian petroleum installations, having lost many such orders to Korean and other Asian yards in recent years. A debate has evolved over the merits of placing such orders overseas, triggered in part by observations of large cost overruns, delays and difficulties meeting Norwegian required standards in several projects.

2.2.3   Fisheries

Fisheries provided 11 per cent of Norway's export value in 2017,[6] making it the second leading export sector after petroleum. It is a small employer on national level (<1%), but remains an important source of employment in many coastal regions. Fisheries encompasses both ocean-going, industrial-scale vessels and small vessels manned by a crew of one or two. Since the 1980s aquaculture – mainly of salmon – has emerged as an important component of the sector.

Employment in fisheries has decreased sharply since the 1950s, driven by a combination of improved equipment and resource limitations. Combined small-scale fishing and farming had for centuries been a common lifestyle along the coast, but largely gave way to modern, specialized employment in the second half of the twentieth century. Former fishermen provided a significant source of able workers for the petroleum sector and its associated maritime operations.

The dangers of depleting important fish stocks became manifest in the 1960s. Regulatory actions were taken to bring this under control, including extensions of national jurisdiction and licensing for fisheries in oceans outside Norway. Fishing permits for other nations' vessels are negotiated. The resource situation for fisheries in Norwegian waters is currently assessed to be largely stable and sustainable, but overfishing remains a major problem for many fish stocks internationally.

2.2.4   Hydroelectricity and Metals

Mountains, rain and snow are prominent features of Norway's geography. They combine to provide excellent resources for hydroelectric power. This was extensively developed in the twentieth century, particularly in the 1950s–80s. Norway's hydroelectricity production averaged 138 terrawatt hours (TWh) per year[7] during 2013–17. Thermal and wind power contributed another 6 TWh. Supplementing the hydroelectric generation capacity, reservoirs have been built with capacity to store water as potential electricity corresponding to two-thirds of annual consumption. This is critical to cover winter requirements, when power consumption is high and precipitation in mountains falls mainly as snow. Norway has never had nuclear power stations, but has operated two small nuclear reactors for research purposes.

---

[6]   January-November 2017.
[7]   Without deduction of transmission losses and so on, which were 11 TWh.

Norway's hydropower capacity corresponds approximately to the electricity consumption of the Netherlands, a country with a population more than three times the size of Norway's. Electricity supply is thus clearly oversized compared to what would have been normal requirements if the country had been using energy in a similar manner as other countries in Northern Europe. In recent years large transmission cables have been built which allow for a significant foreign trade in electricity, but with few exceptions annual net foreign trade in electricity has been less than 10 per cent of production. The development of Norway's large hydroelectric resources therefore depended on making profitable use within the country of the large electricity production.

This has been accomplished in two main ways: by developing power-intensive industries and by utilizing electricity for requirements which would otherwise be met by other fuels. The latter approach is evident as Norwegian homes and other buildings are predominantly heated by electricity, whereas the country's European neighbours tend to use natural gas, district heating or petroleum for such purposes.

Power-intensive industries have been developed in the form of metals, mainly aluminium. Such industries were the direct cause of a number of hydro-electric developments. Norway has no significant deposits of ores containing aluminium, which are therefore imported as bauxite or semi processed oxide. Processing these to obtain metallic aluminium requires large amounts of electricity and abundant hydroelectricity was the decisive reason for developing this industry in Norway. It is strongly exposed to international business cycles and provided around 10 per cent of Norway's non-petroleum export value over the 20 years to 2017.

The significant hydroelectric capacities and associated metal industries have been crucial to the development of the petroleum sector in several ways. The need for concrete dams to form water reservoirs gave rise to competencies in large concrete constructions, which were applied to build the 'Condeep' substructures for several offshore platforms in the 1970s and 1980s. The aluminium industry contributed to building competencies in industrial processing in the country, and one large firm with a background in power-based industries also ventured into becoming a petroleum field operator (Norsk Hydro). On the other hand, the abundance of electricity in the country presented an obstacle to investment in natural gas infrastructure. Consequently, Norway makes little use of natural gas domestically, but has become a large supplier of gas into the highly developed European natural gas pipeline grids.

The regulatory framework for hydroelectric resources reflects the premise that the resources fundamentally belong to Norwegian society, rather than individual property holders. Since 1909 legislation has provided that private parties can acquire a concession and hold rights to utilize hydroelectric resources for a specific period of time (60–80 years), after which they revert to the State. Under current legislation, concessions for acquisition and development of hydropower resources can be acquired only by entities in which there is at least two-thirds public ownership and effective public control.[8] Most hydroelectricity is produced by companies owned by the government and/ or local and regional authorities. Development of large hydropower projects is subject to stringent approval processes and the remaining large undeveloped resources are mostly reserved for natural conservation. Waterfalls with a production capacity of less than 2.9 megawatts[9] do not require a separate concession for the acquisition of ownership, and can be held and developed by private owners. In recent years, small-scale hydro projects have been promoted, which have been developed by farmers and local entrepreneurs with simpler approval processes. Wind power developments, for which Norway has significant potential, have lagged behind those in many other European countries, for reasons which include limitations on transmission capacity and market saturation by hydroelectric production.

2.2.5   Mining

Various metals and minerals have been mined in Norway for centuries. Mining has contributed less than 1 per cent to Norway's GNP over the last 50 years. Coal deposits are present and mined only on the Svalbard islands in the Arctic, where there are Russian as well as Norwegian coalmines due to an international convention allowing for exploitation of resources by several countries.

There are some similarities in legislation for mining and for petroleum. Mining legislation establishes that most metals (but not most non-metallic minerals) are the property of the State, irrespective of the land property where they are found. The present mining law dates from 2009,[10] but has precedents from as far back as the sixteenth century.

2.2.6   Other Non-petroleum Industries

Industrial production was a key driver of Norway's strong economic development from 1945–70, based on strategies of increasing productivity, international competitiveness and closure of unprofitable operations. From the 1970s, industry declined in relative importance to the economy. The contribution made by non-extractive industrial manufacturing to Norway's GNP fell from

---

8   *Vannfallrettighetsloven* (Waterfall Rights Act).
9   The limit is stated in the Act as 4000 horsepower.
10   *Mineralloven* (Minerals Act).

20 per cent in 1970 to 8 per cent in 2015. The petroleum sector, by comparison, grew from 0 to 18 per cent of GNP during the same period.

Norway has never had a viable automobile industry and has few internationally strong brands for consumer goods of any kind. Numerous industrial firms which previously served the home market eventually succumbed to international competition. Many manufacturing operations, particularly labour-intensive operations, have been outsourced to low-cost countries, with the previous Norwegian manufacturer often retaining control over design, branding and marketing. A significant number of food-processing industries continue to operate in Norway, as do several manufacturers of building materials suited to Norwegian building requirements. Norway had sizeable industries for paper and other wood-based products throughout much of the twentieth century. The paper industry in particular has been much diminished, due to stagnant global paper consumption and more cost-effective international competition.

2.2.7   Services

The services sectors delivered 64 per cent of GNP in 2017, up from 61 per cent in 1970.[11] This very large and diverse group of economic activities comprises public services including defence as well as private services, with public services representing about one-third of total services. It mostly serves the domestic population, although some sectors also face international competition (eg, tourism, IT services and airlines). The growing contribution of services to GNP from 1970–2017 occurred mainly in health and social services, which increased from 4 per cent to 11 per cent of GNP. The services sectors in Norway are comparatively large, well developed and automated to reduce the need for labour input, but structurally are not remarkably different from those found in other European countries. A small number of Norwegian services firms have established a significant international presence.

2.2.8   The Petroleum Sector's Impact on the Norwegian Economy

Oil and gas production in 2017 provided 16 per cent of Norway's GNP, 38 per cent of exports, 21 per cent of public sector revenues but only 2 per cent of employment.[12] Its contribution to GNP and exports peaked in 2008, at 28

---

[11]   Service sectors exclusive of pipeline transportation and international shipping.

[12]   Calculated from the Revised national budget 2017, www.statsbudsjettet.no, last accessed 10 January 2018. Not including the industries providing goods and services for petroleum production.

per cent and 64 per cent respectively, as this was a year of exceptionally high oil prices.

Norway's production of oil and gas peaked in 2004 at 4.5 million barrels of oil equivalents per day, of which 62 per cent was crude oil. By 2017, oil production had fallen and gas production had increased, to a combined level of 9 per cent less than the 2004 peak and with natural gas production being slightly larger than total liquids (oil, condensate and natural gas liquids).[13] Combined oil and gas production rose slightly from 2013–17, following earlier declines. As of late 2017, the Norwegian Petroleum Directorate estimated that combined production would remain close to the 2017 level until the mid-2020s and decline thereafter.[14]

Norwegian petroleum production began in 1971 as very modest test production. Substantial volumes started coming in 1975. Since then, the petroleum sector has allowed Norway to maintain higher economic growth and lower unemployment than most other European countries. The industry has supported national employment both directly by its demand for labour (including in sectors supplying goods and services to petroleum operations), and indirectly by permitting a relatively expansive financial policy.

Oil prices increased sharply in 1974 and again in 1980. The early 1980s were consequently a period of high activity and bold project decisions. The sharp drop in oil prices in 1986 had a decidedly sobering impact on the sector. It also triggered a serious downturn in the national economy, during which most banks had to be taken over by the government to prevent their collapse (they were later re-privatized, which allowed the government to recover its expenditures). The 1990s saw cautious growth in the petroleum sector on expectation of oil prices in the USD 10–20 range, leading to the 2004 production peak.

Norway has a sovereign wealth fund, into which the State's net proceeds from petroleum operations are deposited. In January 2018 the fund had accumulated a value of USD 1050 billion, corresponding to more than two years of gross national income. The fund's value in Norwegian currency is continuously updated on the website of the central bank,[15] where information on the fund's management can also be found. The fund was formally established by Parliament in 1990 and received significant deposits from 1996. This was 25 years after the first Norwegian oil production, during which time the government had used the proceeds to pay for its own direct investments in the sector, to pay off debts and for general budgetary support. The fund is

---

[13]   Norwegian Petroleum Directorate, http://factpages.npd.no, last accessed 10 January 2018.

[14]   Nyland, Bente (Petroleum Director), Presentation at Operators' Conference, Stavanger, 15 November 2017, www.npd.no, last accessed 10 January 2018.

[15]   www.norges-bank.no and www.nbim.no, last accessed 10 January 2018.

managed by a department of the central bank. It is invested exclusively outside Norway in a mandated combination of different asset classes. Parliament has established a rule, which is generally respected by most political parties, on how money may be withdrawn from the fund for budgetary support. This should not exceed the anticipated inflation-adjusted investment income of the fund, somewhat optimistically assumed to be 4 per cent per annum, but withdrawals have generally been less than this. The fund's formal name translates to 'Government Pension Fund Global', which is intended to emphasize its purpose as a long-term national savings institution to be shielded from short-term budgetary needs. Another purpose of the fund has been to reduce inflationary pressures which would have resulted from injecting large petroleum proceeds directly into the rather small Norwegian economy.

A salient feature of Norway's petroleum sector is the notable success of the supplier industries. These are businesses which employ Norwegian labour to supply goods and services to petroleum operations not only in Norway, but also worldwide. By one measure, these industries achieved sales of USD 84 billion in 2014, including sales outside Norway of USD 31 billion.[16] These industries are also an important source of employment, exceeding employment directly in petroleum operations. The maritime components of the supplier industries are discussed above in relation to the shipping and shipbuilding sectors. The term 'local content', which is often used internationally to refer to efforts to involve host nation workers and businesses in the sector, can mislead by creating an impression of economic autarky as a goal for petroleum operations. In contrast, these industries have evolved in Norway based on recognition of their international character. In many cases they cannot aim for success in Norway's petroleum sector alone, but must succeed in a global market, if at all. Foreign investment and ownership contribute to Norwegian value creation when substantial operations are undertaken in Norway by Norwegians.

A more complex issue is whether Norway's petroleum sector has caused an erosion of other sectors of the economy, similar to the phenomenon sometimes referred to as the 'Dutch disease'. This would be the case if petroleum activities had caused the currency to strengthen, and prices and other costs to increase to the point where they eroded the international competitiveness of non-petroleum industries. This issue has been of concern to policy makers since the 1970s and is one reason for the establishment of the Government Pension Fund Global. No attempt is made here to fully assess Norway's affliction by the 'Dutch disease'. Several statistical comparisons made in

---

[16] Rystad Energy, *'Internasjonal omsetning fra norske oljeserviceselskaper'* ('International sales by Norwegian petroleum service firms'), Report to the Ministry of Petroleum and Energy, 15 December 2015.

this chapter of Norway's economy in 1970 and in 2017 suggest that Norway can hardly claim to be unaffected by the phenomenon. Salaries and benefits have clearly been more attractive for personnel in petroleum operations than in most other sectors, obviously stimulated by the 78 per cent marginal tax rate which effectively transfers the larger part of such costs to the State. This may have contributed to drive up labour costs for other Norwegian industries, diminishing their international competitiveness. Yet the country retains several non-petroleum business sectors and firms which continue to demonstrate the capacity to prevail in international competition.

A test of the economy's resilience to changing petroleum sector fortunes came with the 2014 fall in oil prices. Unemployment, which had stood at 3.5 per cent in 2013, rose to a peak of 4.8 per cent in August 2016, as many firms supplying goods and services to petroleum operations reduced staffing. Job losses in the sector were partly offset by expansive fiscal policy and public expenditure, which the government was in a position to comfortably afford. A weakened currency as a result of the lower oil prices aided some non-petroleum sectors exposed to international competition. Unemployment fell to 4.0 per cent in late 2017, also aided by renewed confidence in the petro-leum sector as oil prices rose above USD 60 per barrel.

2.3   EMPLOYMENT AND HUMAN RESOURCES

2.3.1   Employment Structure

Table 2.1 summarizes employment in total and by sector in 1970 and in 2017.

This table describes Norway as of 2017 as a post-industrial economy dom-inated by services sectors. Compared to 1970, the share of the population in gainful employment has increased from 38 per cent to 50 per cent, due to the increased participation of women in the workforce.

Farming, forests and fisheries had employed more than 30 per cent of the workforce after the Second World War, but were down to 11 per cent in 1970 and 2 per cent in 2017; while improved equipment allowed the amounts of foods produced from land and sea to be largely upheld by the diminished workforces.

Manufacturing industries in 1970 were near their peak in terms of employ-ment and relative contribution to Norway's GNP. Productivity gains from 1945–70 had largely resulted in production growth, but after 1970 continued productivity gains with less production growth inevitably led to reduced employment. Norway's remaining industries as of 2017 tend to be highly auto-mated, capital intensive and, where manual work processes are still needed, partially outsourced to low-cost countries. Of the 10 per cent of employment still provided by industries, about one-third is in the petroleum sector and

Table 2.1   *Employment in Norway, 1970 and 2017 (annual averages)*

Year	1970	2017*
Population total	3 876 000	5 271 000
Persons in employment, including self-employment	1 488 000	2 647 000
of whom: Men (%)	71	53
Women (%)	29	47
Unemployment *(job seekers' share of work force)* (%)	1.7	4.4
Share of persons employed, by sector (%)		
Agriculture, forestry and fisheries	11	2
Industries (manufacturing and extractive)	28	10
Maritime transportation	4	1
Construction and utilities	10	9
Trade, retail and finance	17	16
Public authorities and defence	5	7
Educational services	5	8
Health and social services	6	21
Other services	15	26

Note: * Four quarters ending Q3 2017.
Source: Statistics Norway.

its supplier industries, which were virtually non-existent in 1970. The shipping industry (maritime transportation), while still a formidable Norwegian business, employs few Norwegians now, but larger numbers of sailors from countries where lower salary levels are accepted.

Employment growth has occurred in the services sectors, notably health and education, which are largely (but not exclusively) provided by the public sector. The strong employment growth in health and social services is in part a consequence of the increased formal employment of women. Care for the elderly and others has largely shifted from being provided by female family members outside formal employment to being provided by women (and fewer men) employed in health service organizations.

2.3.2   Education

In 1970, 7 per cent of the adult population had completed an officially recognized education at university level. By 2016, this figure had risen to 33 per cent.

Since the 1980s more women than men have completed university-level education. In the age range of 30–39 years in 2016, 57 per cent of women and 40 per cent of men had completed a degree at university level. This includes 19 per cent of women and 17 per cent of men having completed university degrees of more than four scheduled years. There are significant differences in the educational choices of men and women: there were 71 per cent women among those with a university level degree in health and social professions, but 71 per cent men among those with a university degree in technical professions and natural sciences.

Some 39 per cent of men and 26 per cent of women in the age range of 30–39 years (as of 2016) had ended their formal education with a secondary school degree. A large majority of these – five out of six – had trained for a vocational profession. Vocational training at the secondary level (normally 16–19-year-olds) involves a combination of school attendance and supervised work in a business, leading to a certificate of apprenticeship ('*fagbrev*'). Continuing educational opportunities are available to those who have attained this qualification. Vocational training is available for a wide range of professions, with the largest numbers of graduates in technical professions including construction (45 per cent), health and social professions (17 per cent) and administrative professions (11 per cent). Again, there are significant differences in the choices of men and women: there were 90 per cent men among those qualified for the technical and construction related professions, and 85 per cent women for the health and social professions.

2.3.3   Time at Work, or Not at Work

In the first half of the twentieth century, young men were normally expected to seek employment from the age of around 15. Many were hired as sailors in the merchant fleet even at such young age. Access to (and the attraction of) advanced education changed this gradually during the second half of the century, increasing the age at which young people enter the workforce in full employment. Part-time work and work during university vacations have become quite common among students. In 2015, 28 per cent of persons aged 15–24 years were in part-time employment, and 22 per cent were in full-time employment. Men in that age group were more often in full-time employment, women more often worked part time and 50 per cent of both genders were in some degree of employment.

Military service has been one factor that has kept young men away from the labour market for some time. Norway has a mandatory military defence obligation for men and, since 2015, also for women. For several decades following the Second World War, most able-bodied men at the age of 19 were called up, usually for 12 months of initial service and then shorter manoeuvres of three weeks a couple of times later in life. For reasons which include the degree of professionalism required with modern warfare technologies, the armed forces no longer wish to draft large annual contingents for conventional combat training. As a result, the forces have become increasingly selective in admitting only those who demonstrate capabilities and motivation for service. This has for some time included a number of young women on a voluntary basis.

Since the 1970s, Norway has had more generous benefits for childbirth than many other Western countries. As of 2017, childbirth entitles parents to a total of 49 weeks of paid parental leave, provided that the father takes at least 10 weeks. Long parental leave is often cited as a reason for the high participation of young women in the workforce.

The average working week in 2016 was 34 hours, taking account of overtime and part-time work. Normal working conditions are 37.5 hours per week and five weeks of vacation per year (six weeks for those aged 60 and above). There are statutory limits on the amount of overtime permitted. There are 10 days of public holidays per year, some of which may coincide with weekends. Those required to work outside normal business hours usually have fewer working hours, additional vacation and/or earlier retirement.

Part-time work is more common among women than men. Thus, in 2015, 54 per cent of the persons in employment were men, but they accounted for 58 per cent of the time worked. Part-time work is common in the public health and social sector, where 80 per cent of employees are women. For this and other reasons, this sector accounted for 20 per cent of the employed population in 2015, but only 17 per cent of the work provided.

Unemployment has generally been lower in Norway than in most other Western European countries. From 1999–2016, the average unemployment rate was 2.9 per cent. The 2017 average rate of 4.4 per cent is high by Norwegian standards, but half of the euro area average. Policies have put a high priority on avoiding unemployment and strong public finances have made such policies affordable.

A greater concern for Norway's workforce has been the high levels of absence for sickness and disability. In 2017, 6.5 per cent of employees were absent for sickness on average;[17] while 14 per cent of the population aged 18–67 received disability pension or related support, in most cases for full incapability.[18] These shares of the working-age population not working due

---

[17] Norwegian Directorate for Labour and Social Security (NAV), *Sykefraværsnotat 3. Kvartal* 2017 (Statistical note on absence for sickness, 3rd quarter, 2017), www.nav .no/no/NAV+og+samfunn/Statistikk/Sykefravar+-+statistikk/Sykefravar, last accessed 10 January 2018.

[18] This 14 per cent comprises two categories: disability pension (9.6 per cent) and work assessment allowance (4.1 per cent). Data from NAV: *Utviklingen i uføretrygd*

to sickness or disability have been trending slightly downwards over the last several years. One explanation sometimes offered for the high rates of compensated non-work is that they may relate to the high rate of general participation in the workforce.

Norwegians born in 2017 were expected to live to age 82 on average, whereas those who reached the age of 60 in 2017 were expected to live to nearly 85. The standard retirement age has long been 67. Since 2011, there has been a flexible retirement system allowing for retirement at age 62 or later, but with higher pension payments from the national pension system for those who opt for later retirement or partial pension payments in the early years of retirement. Combinations of work and pension payments are possible. In 2017, 32 per cent of those aged 62–66 received a retirement pension from the social security system, but 57 per cent of these continued to earn income from work while also receiving pension payments.[19] Employees may opt to keep working beyond age 67 until they reach the employer's standard retirement age, usually 70 years. In the absence of this standard retirement age, an employer is entitled to terminate employment due to age only when the employee has reached age 72 (recently increased from 70). In 2017, 16 per cent of those aged 67–69 and 7 per cent of those aged 70–74 received income from work in addition to their pensions.[20]

2.4   DEMOGRAPHICS

2.4.1   Population Increase and Fertility

Norway's population grew from 3.87 million at the beginning of 1970 to 5.26 million at the beginning of 2017. The average growth rate was 0.65 per cent per year. All years in that period saw positive population growth. The growth over that period came almost equally from excess births over deaths and from net immigration.

---

per 30. september 2017 (Statistical note on disability pensions as of 30 September 2017), and *Personer med nedsatt arbeidsevne og personer med rett til arbeidsavklaringspenger, September 2017* (Persons with reduced work capacity and persons entitled to work assessment allowance, September 2017), www.nav.no/no/NAV+og+samfunn/Statistikk/AAP+nedsatt+arbeidsevne+og+uforetrygd+-+statistikk, last accessed 10 January 2018.

[19]   NAV, *Utviklingen i alderspensjon per 30. september 2017* (Statistical note on old age pensions as of 30 September 2017), www.nav.no/no/NAV+og+samfunn/Statistikk/Pensjon+-+statistikk, last accessed 10 January 2018.

[20]   See n 19.

In 2016, the fertility rate for Norwegian women was 1.71, being the average number of children anticipated to be born to each woman during her lifetime. The average during 1990–2016 was 1.85, which was higher than the majority of European countries. In the nineteenth century the Norwegian fertility rate was above 4 – similar to many African countries presently; but it fell in the first four decades of the twentieth century as the country became increasingly industrialized and affected by international recession in the 1930s. After a post-war period of fertility rates between 2 and 3, the rate fell to less than 1.7 in the 1980s.

The somewhat higher fertility rate since the 1990s has been attributed to public measures designed to make it easier for women to have a working life while also having children: extensive parental leave, availability of kindergartens and legislation to prevent discrimination against women at work. Studies by researchers at Statistics Norway have shown a reduction in the earlier tendency for highly educated women to have no or few children. They have also analysed the components of the more recent (2008–14) reduction in fertility. Women recently have tended to wait longer before having their first child: this occurred at the average age of 28.7 years in 2014, an increase from 25.5 in 1990, with a significant part of the increase in recent years. There is also a reduced tendency for women to have three or more children, especially among women with little education.[21]

2.4.2   Immigration

In early 2017, 13.8 per cent of the resident population of Norway were immigrants, compared to 1.5 per cent in 1970.[22] In 2017, another 3.0 per cent had been born in Norway of immigrant parents, and 5.6 per cent had one immigrant and one Norwegian-born parent. There was net immigration to Norway in all years from 1967–2015.

During 1967–2016, Norway had net immigration of 695 000 persons. The largest group, 107 000, came from Poland. Some 90 per cent of these Poles, as well as significant numbers from other Central and East European countries now in the EU, have come since 2005. The percentage of immigrants in Norway has doubled since 2005, mainly due to immigrants from Europe, but also significant numbers from war-torn or repressive countries in Asia and Africa. Syrians were the largest immigrant group in 2016. The immigrants

[21]   Trude Lappegård and Lars Dommermuth, '*Hvorfor faller frutkbarheten i Norge?*' ('Why is fertility declining in Norway?') (Statistics Norway, Økonomiske analyser 4/2015).

[22]   An 'immigrant' is defined as a person born as a foreign resident and having become a resident of Norway.

from Poland and other recent EU members clearly came to seek work during a period when Norway's economy ran at stretched capacity under the influence of high oil prices, and became an important supplement to the workforce, especially in the construction sector. As the economy turned and overcapacities emerged in 2015, net immigration continued at a somewhat reduced rate.

Norway has a longstanding arrangement with its Nordic neighbours, making these countries effectively a common labour market. There has for many decades been significant migration within the region, helped by the fact that the Swedish, Danish and Norwegian languages are mutually comprehensible (with a bit of goodwill and practice). Since 2005 there has been significant net immigration into Norway from all other Nordic countries. Before that, the net flows have shifted over time. During 2015–16 there was again net emigration from Norway to the other Nordic countries, as Norway's economy was hit by lower oil prices.

Restrictions on immigration were imposed in 1975, with the intent of shielding the Norwegian labour market from a large influx of jobseekers. This was driven mainly by a general concern about potential unemployment at that time, and not by any experience of large immigration to Norway. Residence and work permits were (and are still) granted to specialists whose skills were needed, which was important for the emerging petroleum industry at the time. Other than this, immigration is allowed as a consequence of Norway's agreements with the European Union, including the Schengen Agreement, as well as admissions of refugees, asylum seekers and families seeking reunification.

The large influx of refugees and migrants in 2015 triggered a tightening of the rules for refugees and asylum in Norway, as in several other European countries. The issue received much attention when surprisingly large numbers entered Norway at its border with Russia in the far North, coming on bicycles due to a rule against crossing the border on foot. In total, the Norwegian Directorate of Immigration has reported[23] that 31 145 persons arrived in Norway to apply for asylum in 2015. The flow was much reduced in late 2015.

2.4.3   Diversity and Homogeneity

Putnam (2007) has argued that the increasing ethnic diversity of a society, which is a consequence of increased immigration to European countries, challenges social solidarity and inhibits social capital in the short to medium term, while creating desirable new forms of social solidarity in the medium to long

---

[23]   www.udi.no, last accessed 10 January 2018.

term.[24] If Putnam is right, this has important implications for Norway, as it is changing from a quite homogeneous to a more diverse society.

Norway has traditionally been perceived as a quite homogeneous society, culturally and ethnically. Before immigration in the 1970s, the only significant national minority was the culturally distinct Sami people, which represented less than 1 per cent of Norway's population. The Sami's traditional homelands since prehistoric times have been Northern Norway and adjacent parts of Sweden, Finland and Russia. Descendants from small groups of Jews, Roma and Finns who immigrated over the last several centuries have preserved some cultural distinctiveness and are recognized as national minorities, but their numbers are apparently too small to have altered the general impression of Norway's cultural and ethnic homogeneity.

Immigration over the last several decades has established a significant presence in Norway of people who, in cultural and ethnical terms, are distinctly different from the indigenous Norwegian population. Pakistanis and Vietnamese arrived from the 1970s onwards, followed by several other groups from Asia and Africa. The homogeneous society of the 1960s and before is giving way to a society in which different groups of people live side by side in observation of different cultural norms.

2.4.4   Spatial Population Structure

Norway is divided into 19 counties (*fylker*) and 428 municipalities (*kommuner*) as of 2017. Their numbers will be reduced to 11 and 356, respectively, from 2020. The outlying islands of Svalbard and a few others are subject to different administrative arrangements, as it is assumed that the people staying there do so for limited periods and have their permanent residences elsewhere.

Norway's population density, at 13.5 per square kilometre, is the lowest on the European mainland. Population density is highest in and around Oslo, and is lowest in the Northern counties, as well as the mountain regions in Southern Norway.

Oslo is the capital city and the most populous municipality, with a population of 672 062 as of 30 September 2017, which was 12.7 per cent of the national population. The ten largest municipalities had 34 per cent of the national population as of 1 January 2016. The same municipalities (including previously separate entities which have been merged into them) had 28 per cent of the national population in 1970 and 25 per cent in 1951. The increased population share living in a few large municipalities indicates a tendency of

---

[24]   Robert D. Putnam, 'E Pluribus Unum: Diversity and Community in the Twenty-first Century', *Scandinavian Political Studies*, Vol 30 – No 2, 2007, pp137–174.

concentration, as is also observed in many other countries. The regions attracting most population growth in the second half of the twentieth century were around Oslo and regions in Southwest Norway, where much of the petroleum industry has developed.

The much reduced employment in agriculture, forestry and fisheries has triggered changes to where people choose to live. Many municipalities have indeed seen a reduction in population since 1951 or since 1970. However, successive governments have pursued policies of maintaining population in all parts of the country, if not necessarily in every local community. All 19 counties saw population increases from 1951–70, and with two slight exceptions all of them also saw population increases from 1970–2016. People have moved from small agricultural or fishing communities not only to the large cities, but also to growing communities closer to their previous homes, where opportunities in education and a variety of modern jobs have been available.

The largest relocations of people within Norway occurred in the 1950s–80s and have become less intense in recent years. The increased employment of women, along with the fact that most Norwegians own their homes, now tends to create higher hurdles for established families to move long distances.

2.5   INFRASTRUCTURE AND ACCESS TO MARKETS

2.5.1   Physical Infrastructure and Transportation

Developing the road network was a major public undertaking throughout the twentieth century. The nature of the road network reflects the country's rugged terrain and sparse population. Roads now connect all significant settlements, with the exceptions of some remote islands and other locations where few people live; such remaining road-less settlements are usually reached by public boat services. Road standards are not particularly high in comparison to other European countries. Four-lane (or larger) freeways are found only near the largest cities, not as long-distance connections. Most road traffic drives on two-lane and often rather winding roads. Driving along the Western and Northern Coasts involves a number of ferry crossings. The main roads are kept open throughout the year, but there are usually a few incidents each winter when the roads between Eastern and Western Norway, and some in the North, are closed or require convoy driving due to storms in the mountains.

Reducing traffic accidents has been a high priority of Norwegian policy. After a peak of 560 road fatalities in 1970, this figure was reduced to 107 in 2017 in spite of increased traffic volume.[25] Norway has had the lowest per-capita number of road fatalities in Europe in recent years.

Some 269 million tonnes of cargo were moved on Norwegian roads in 2016, of which 264 million tonnes were inland transports. In terms of inland transport work (tonnes x kilometres of goods moved within Norway), ships carried 48 per cent, road vehicles also 48 per cent and rail 5 per cent. In addition, transports by sea to and from offshore petroleum installations corresponded to 74 per cent of the inland transport work. Air cargo transport is negligible as a share of tonnes moved, but is important for mail service and express deliveries.

Air travel is much used in Norway. In 2017 there were about 50 airports served by scheduled flights, of which eight had daily international departures. Many small airports were built in the 1960s–80s, serving communities in the Western and Northern regions. The busiest routes between the larger cities are usually served by two competing carriers. The small airports are served mainly by one carrier under a public contract. A total 27.3 million passengers boarded an aircraft in Norway in 2016, 41 per cent of which were international departures.

Railways have a relatively modest role in the Norwegian infrastructure system. The railways were developed mainly in the first half of the twentieth century and have seen only limited expansion since. There are railways between Oslo and the major towns (except Tromsø in the North), but with significant capacity limitations. Long-distance rail connections are mostly single-track, being used for passenger traffic as well as containerized goods. The only high-speed train connection is between central Oslo and Oslo Airport at Gardermoen. The northern termination of the Norwegian rail system is at Bodø. Further north there is a railway to Narvik from Sweden, used primarily for exporting Swedish iron ore. In 2014, 74 million passenger journeys by rail were made and 34 million tonnes of cargo were moved on Norwegian rail, of which more than half comprised Swedish iron ore in transit through Narvik.

Norway's coast is well suited for shipping and ports. Coastal and North Sea shipping on modestly sized vessels handles much of the bulk and containerized transport requirements. There are numerous ports serving local communities and businesses, and ports serving the oil industry for supplies and for exports of oil and liquefied gases. Apart from the Swedish iron ore shipped out of Narvik, Norwegian ports are not much used for international transit, as they are generally too small to handle the largest container ships (the nearest large container port is in Gothenburg, Sweden). Passenger transport along the coast has been in relative decline in recent decades, being replaced by air and road

---

[25] www.vegvesen.no, last accessed 10 January 2018.

transport, but high-speed boats still serve many coastal and island communities. An enduring institution are the daily departures between Bergen and
Kirkenes (on the Russian border in the North) of combined passenger and
general cargo ships ('*Hurtigruten*'), which increasingly cater to tourists, while
remaining an essential transport service for many communities.

2.5.2   Access to Export Markets

Norway is not a member of the EU, having twice held referenda in which
voters turned down accession agreements which the government had negotiated with the EU. In 1960 Norway joined the European Free Trade Association
(EFTA), the European free trade bloc whose largest member was the UK
until the latter joined the European Union in 1973. The EFTA still exists,
with Norway as one of the few remaining members. More importantly, after
Norway turned down EU membership in 1994, its relationship with the EU
was based on the framework of the European Economic Area (EEA). The
EEA Agreement[26] provides for Norway to be part of the EU internal market
by incorporating the EU's internal market legislation, including energy market
directives, into Norwegian law. This provides for similar market integration
as if Norway were a member of the EU, but without certain obligations and
without the voting influence in the EU. Norway is part of the Schengen Accord
for unrestrained movement of people across borders, which has triggered significant immigration of East Europeans in search of jobs. Norway has joined
the European system for climate-related emission rights. Norway has retained
its own national currency.

   All Norwegian governments since 1945 have pursued policies of facilitating
international trade for a wide range of goods and services, which has been
important for Norway as a small and open economy ambitious for growing
affluence. Industrial goods are traded between Norway and other countries
with few restrictions. An exception concerns the agricultural sector, which
remains strongly protected by import tariffs, restrictions and subsidies in order
to sustain domestic food production and avoid de-population of parts of the
country. Agriculture and fisheries were not included in the EEA Agreement,
for rather different reasons: Norway wanted to shield its agriculture from
foreign competition in order to maintain settlements throughout the country.
Norway also wanted to maintain full control over its fish resources, while still
seeking access to the EU as a market for fish exports. Fish exports have met

---

[26]   The Agreement on the European Economic Area covers the EU member states,
Norway, Iceland and Liechtenstein; www.efta.int/Legal-Text/EEA-Agreement-1327,
last accessed 10 January 2018.

barriers to trade at times, such as a penalty tariff imposed by the US government on salmon imports from Norway during 1991–2011. Norway has protected its fisheries by restricting foreign vessels' access to fish in Norwegian waters.

Prior to the EEA since 1992, Norway had several arrangements with its Nordic neighbours and other parts of Europe to facilitate trade. Norway has been a member of the World Trade Organization and its precursor, the General Agreement on Tariffs and Trade, since its foundation in 1947.

2.6 WELFARE SYSTEMS

Norway has extensive public welfare systems, comprising universal welfare rights available to all citizens. They can be grouped broadly into two categories: transfers and services. There follows a summary of key welfare transfers and social services.

2.6.1 Welfare Transfers

Welfare transfers comprise pensions for age and disability, unemployment benefit, child support, childbirth benefit, compensated sickness leave, grants for education, grants for certain other circumstances and social assistance. With the exception of the latter, welfare transfers are entitlements triggered by the circumstances of the recipient. Social assistance is intended as a temporary support for persons unable to find sufficient means for proper subsistence.

Reforms to the Norwegian pension system were implemented in 2011, in response to challenges similar to those encountered in most other Western economies: increasing longevity and a declining ratio of people in employment to people in retirement. As the standard retirement age had been 67 years since 1973 (albeit with early retirement options for a number of professions), Norway has had a tradition of retirement at a higher age than many other European countries. Following the 2011 reform, retirement at 62 is an option for all, but with economic incentives for working after that age and with a right to work until at least age 70 (72 since 2015) in most professions. The pension reform was passed with support from most parties in Parliament and without major public controversy.

The public pension system is supplemented by pensions funded by contributions from employers. Such contributions became mandatory at certain levels from 2006 and were common on a voluntary or negotiated basis before that.

As discussed in section 2.1.2, Norway has comparatively generous benefits for disability and sickness, and the proportion of the workforce absent for those reasons has sometimes been described as alarmingly high. Reductions in sickness benefit would be politically controversial. Several attempts have been made through other means to contain sickness absences and engage people on disability pensions in work. Some reductions in sickness absences have been achieved: this figure peaked at 8.2 per cent in 2003 and had fallen to 6.5 per cent by 2017, but seems to have levelled out more recently.

Unemployment benefit is normally about 62 per cent of recent employment income, with certain limitations, and limited in time to two years. Restrictions against employers laying off staff in times of slack business are less severe in Norway than in a number of other European countries, and companies can generally do this if necessary. It is also possible to lay off staff temporarily, in which case they receive unemployment benefit instead of salary but are entitled to have their job back when and if the firm is in a position to resume its previous staffing levels. The recipients of unemployment benefits are offered training opportunities and are required to seek work actively.

Parents receive NOK 970 per month per child below the age of 18 (2016). Single parents receive higher amounts. In addition, there is an arrangement for financial support for small children who are not in kindergarten. This support was intended to compensate parents who do not benefit from the public kindergarten subsidy and give parents a more balanced choice as to whether to take their children to kindergarten. While the general child support has been largely uncontroversial, the additional support for children not in kindergarten has been politically controversial. Critics argue that keeping children away from kindergarten will slow their social integration (including the cultural integration of immigrant children) and dissuade mothers from pursuing a working career.

Childbirth triggers an entitlement to 100 per cent salary compensation for 49 weeks or 80 per cent compensation for 59 weeks, during which time one parent stays away from work. There is a requirement that each parent take at least 10 weeks of paid leave or lose that portion of the benefit. This requirement is motivated by a political desire to engage fathers fully in caring for small children and to promote equal career opportunities for both genders.

Young adults pursuing a recognized education are entitled to educational support intended for their subsistence while studying. Most of this is in the form of loans on favourable terms, including conversion of parts of the loans to a grant upon the successful completion of education. Support is also available to study outside Norway, which is quite common among young Norwegians. Many students supplement the money from the State with some income from part-time jobs (or from generous parents) in order to improve their standard of living while studying.

Social assistance is intended for persons who lack other means of adequate subsistence. It was granted to 130 000 persons in 2016, at a total of NOK 6.2 billion. Earlier statistics have indicated that immigrants comprise a significantly higher share of those receiving social assistance than their share of the general population. Young adults aged 20–24 are also relatively numerous among recipients, reflecting that some in this age group have not yet been properly employed and therefore cannot claim unemployment benefit. The overall number of recipients has somewhat reduced from a peak in the 1990s.

2.6.2   Social Services

Most but not all hospitals, care homes, universities, schools and kindergartens in Norway are owned and run by public entities, charging no or heavily subsidized fees to their users. Among these, kindergartens tend to be the costliest for users, with a maximum rate of NOK 2910 per month (in 2018) applied in most locations. This is still a significantly subsidized rate, as kindergartens are extensively staffed and aim to give children an educational experience.

Medical doctors in private practice usually have an agreement with the public sector which allows them to see patients at subsidized fee rates. Prescribed medication is also available at subsidized costs. People with disabilities can usually get equipment and, in some cases, extensive personal assistance for the practical challenges in life. Following a reform in 1991, persons with mental disabilities are no longer placed in large institutions, but usually in smaller homes or other facilities near their place of origin.

Some private hospitals and medical clinics offer medical services outside the public health system, often motivated by some patients' desire to bypass the sometimes rather lengthy waiting times for public services. The costs of the private services are then paid directly by patients or by insurance usually provided by employers.

Standard dental services are paid for directly, with no or little public subsidy, by most adult patients. Children and young adults receive heavily subsidized dental care.

2.6.3   International Comparisons

At USD 30 263 per capita, Norwegian government expenditures were 63 per cent higher than the OECD average in 2015. As a percentage of GDP, this was 48.8 per cent for Norway versus 45.3 per cent for the OECD.[27]

Norway spent the equivalent of USD 6190 per capita on health services in 2015, 85 per cent of which was spent on public health services. The public spending per capita was the highest of all OECD countries except Luxembourg.

---

[27] Calculated from the OECD database, https://data.oecd.org, last accessed 10 January 2018; likewise for the data on health and educational expenditures in the following two paragraphs.

The total spending per capita was 35 per cent less in Norway than the United States, but clearly higher than most other OECD countries.

Norway spent 6.2 per cent of GDP on education (all levels to tertiary) in 2014. The United Kingdom topped this score within the OECD, with 6.6 per cent; most other members scored lower than Norway. Some 43 per cent of Norway's population aged 25–64 years have completed tertiary education, which is in the highest quartile of OECD member countries.

As an indicator of what has been achieved through the resources spent on public services, Norway ranked first on the United Nations Development Programme Human Development Index in 2016, as it has for several recent years.[28] It ranked sixth in 1990, when this index was first published. The index is composed of several indicators covering health, educational, economic and other conditions influencing the standards of human life in 188 countries.

2.7 CULTURAL PRACTICES AND TRANSPARENCY

Hofstede (1991) introduced four dimensions through which to characterize national cultures and conducted surveys based on which a number of national cultures in diverse parts of the world were charted according to these four dimensions. In the updated 2010 edition of the book, two more dimensions were added and Norway's culture was assessed.[29] Norway is described as follows, where the index values (in brackets) range from 0–100. The index values have the presumable advantage of having been established by research-ers outside Norway and are clearly open to debate, but may seem not entirely unreasonable to someone from the country. They are also broadly similar to the assessments for Norway's Nordic neighbours:

- Power distance (31): Low power distance. Hierarchy is for convenience only, relations are informal, communication is direct and participative.
- Individualism (69): Individualistic culture. 'Self' is important. Personal, individual opinions are valued. Privacy is important. Clear distinctions between work and private life.
- Masculinity (8): Strongly feminine culture. Emphasis on 'soft' aspects of life: wellbeing, empathy, solidarity and care for the environment; less on competitiveness and winning.
- Uncertainty avoidance (50): Norway shows no clear cultural distinc-tion with respect to accepting or feeling threatened by ambiguity and uncertainty.

---

[28] United Nations Development Programme, Human Development Report 2016, hdr.undp.org, last accessed 10 January 2018.

[29] www.geerthofstede.com, last accessed 10 January 2018.

- Long-term orientation (35): A relatively low score, indicating normative thinking, respect for traditions, a small propensity to save for the future and a focus on quick results.
- Indulgence (55): In restrained cultures, children are raised to strongly control their desires and impulses, while indulgent cultures are the opposite. Norway's score shows no clear distinction in either direction.

Letters, reports, preparations and meeting documents from public decision processes are in the public domain and usually accessible online, unless specifically exempt for a valid reason. All joint stock companies (the predominant form for Norwegian businesses with more than one owner) are required to submit their annual accounts for publication, which are readily accessible on a website. Even summary personal tax information is public in principle, but this has been controversial and access has been somewhat restricted (the identity of the inquirer will be revealed to the tax subject in question).

In public decision-making processes, there are requirements that persons recuse themselves if they have a personal interest in the matter at hand. As is common in most Western firms, employees of Norwegian firms are usually required by internal policies to declare or reject any gifts or hospitality from business contacts beyond a symbolic value. Such standards appear to have become more restrictive in recent decades. Complaints of corruption in Norwegian public life are rare, although there have been infrequent cases of abuse of a public position for major personal gain. Somewhat more frequent are cases where a public figure appears to have obtained an illegitimate but small personal advantage, perhaps more out of carelessness than of any malevolent intent. Even such cases tend to be quite damaging for that person's further career.

Norway scored 85 on the Corruption Perception Index in 2016.[30] The index reflects the perceived level of public corruption, with a range from 0 (extremely corrupt) to 100 (extremely non-corrupt). Denmark and New Zealand scored highest with 90; Norway ranked in sixth place. There is no reliable index of actual, rather than perceived, corruption.

An anti-corrupt national culture has not prevented Norwegian firms from becoming involved in cases of apparent corruption on a significant scale abroad. The most prolific cases over the last 20 years have involved firms with a large State ownership. In addition to legal repercussions, these cases have led to resignations of leading personnel at those firms and to parliamentary hearings.

---

[30] Transparency International: Corruption Perception Index 2016, www .transparency.org, last accessed 10 January 2018.

2.8   THE POLITICAL CONTEXT

As of 2018, there are nine political parties in Norway's Parliament. This seemingly fragmented composition has tended to provide a reasonable degree of stability to the political system. Most parties have been around for a long time: 167 out of 169 representatives belong to parties with more than 40 years of history in Parliament. They have traditions of working together, finding compromises and forming coalition governments. Since 1961 there has been no single-party majority in Parliament. Governments since then have been coalitions of more than one party, minority governments depending on support from other parties or both. As the Constitution does not allow for elections outside the regular four-year cycle, the system depends on moderation and pragmatism.

The largest party at most times since 1945 has been the Labour Party, which is of a moderate social democratic orientation. According to the main left-right divide in Norwegian politics, the Labour Party leads the left, which traditionally has been labelled 'socialist', although any direct references to 'socialism' are hard to find in its current policy documents. Its policies emphasize a strong public role in an economic framework which also includes private enterprise. The right side of Norwegian politics is more fragmented, with the moderate Conservative Party usually in a leading role trying to build coalitions with several smaller parties to enable a government. Since 1970, seven of the nine parties in Parliament as of 2018 have had their turn in government.

Al-Kasim (2006) has described the formative processes and circumstances of Norway's management of its petroleum sector from the perspective of someone from a very different (in his case, Iraqi) background who became deeply involved in those processes.[31] In contrast to many other nations that discover the prospect of wealth from oil, Norway's political leadership and the general public displayed neither euphoria nor haste in laying their hands on those riches. There was a fair degree of reluctance, concern for loss of national control and a desire to avoid serious mistakes in the early years around 1970.

This caution was clearly influenced by the pervasive political controversies of the time – above all, whether Norway should join the European Community (later, the EU). This was an issue in which the country's political leadership encountered uncommon and fierce popular protest, based on a broad spectrum of political beliefs, which resulted in a rejection by referendum of the negotiated accession treaty in 1972. As in other parts of Western Europe, as well as North America, this was a time of social and political upheaval, of

---

[31]   Farouk Al-Kasim, *Managing Petroleum Resources: The 'Norwegian Model' in a Broad Perspective* (Oxford Institute of Energy Studies, 2006).

inter-generational conflict and of questioning many beliefs, norms and allegiances which had dominated the post-war period of economic growth and the Cold War. In Norway, this led not only to rejuvenated left-wing politics but also to the formation in 1973 of an anti-tax party with evidently populist and right-wing views, which, with a change of name and of political style, has evolved and joined Norway's government in 2013.

This caution in pursuing petroleum developments paid off in several ways. Subsequent development of the sector was supported by a broad political consensus, even though certain issues were contested. Governments could take the time needed to develop a well-considered framework. And a patient attitude lent strength to Norway's bargaining position with international oil firms – not least for the licences awarded in 1973, which led to the large Statfjord oil discovery in 1974.

2.9   ENVIRONMENTAL STANDARDS

Environmental protection has been pronounced a high priority of all Norwegian governments in recent times. The country has enacted legislation intended to inhibit any activity carrying a significant risk of environmental degradation. There have nevertheless been controversies over what an ambitious environmental policy should provide. On one occasion, a government resigned on the grounds of an environmental controversy involving the construction of a gas power plant (the power plant was eventually built, but was not a commercial success and has since been decommissioned). The petroleum industry is subject to strict environmental protection regulations, to avoid pollution from disasters and ongoing operations, and to restrain emissions of climate gases. The latter is one reason for Norway's early introduction of a prohibition against large-scale flaring of natural gas.

Article 112 of Norway's Constitution recognizes a right for everyone to a healthy environment and the principle of sustaining the natural environment, confirming the responsibility of the State to act appropriately to those ends. The Pollution Control Act (first passed in 1981), the Nature Diversity Act (2009) and the Greenhouse Gas Emission Trading Act (2004) are central pieces of Norwegian legislation dedicated to environmental issues. In addition, environmental considerations are integrated in a large number of other acts. On its website the Norwegian Environment Agency[32] lists a selection of the most

---

[32]   www.miljodirektoratet.no/en/Legislation1/Acts/, last accessed 10 January 2018.

important acts under which it has responsibilities for environmental issues. These include the following:

- Pollution Control Act: Subjects emissions to air, waters and ground to restrictive permits and reporting, and rules on handling waste;
- Nature Diversity Act: Protection of nature, its ecosystems and species through sustainable interaction;
- Greenhouse Gas Emission Trading Act: Norway's participation in the European emission trading system;
- Outdoor Recreation Act: Provides for public access to lands (including most privately held land) for recreational purposes;
- Product Control Act: Addresses safety and environmental issues with marketed products;
- Gene Technology Act: Restricts the production and use of genetically modified organisms;
- Svalbard Environmental Protection Act: Sets out special provisions for the ecologically vulnerable Northern islands; and
- Environmental Information Act: Sets out citizens' rights to information and participation in public decision processes with environmental implications.

Following the Paris Agreement of 12 December 2015,[33] Norway has adopted a Climate Act, which sets targets for reduced emissions affecting the global climate. Other legislation with environmental protection provisions includes the Planning and Construction Act, the Work Environment Act, the Health Protection Act, the Wildlife Act, the Mountains Act, the Energy Act, the Petroleum Act, the Aquaculture Act, the Roads Act and the Natural Monitoring Act.

The Petroleum Act, which is the general legislation governing the petroleum sector, defines the objectives and responsibilities for the environmental management of this sector.

The Norwegian Environment Agency is the national authority dedicated to environmental protection, but environmental considerations are an integral part of duties of all state authorities. The 19 regional authorities and 428 municipalities also have significant responsibilities for environmental protection, not least for spatial planning and construction.

The Norwegian Environment Agency manages a separate website dedicated to assessments of Norway's current environmental conditions.[34] Occurrences of inferior air quality in some cities are noted, especially on cold and calm

---

[33]   http://unfccc.int/paris_agreement/items/9485.php, last accessed 10 January 2018.

[34]   www.miljostatus.no, last accessed 10 January 2018.

winter days when air tends to be immobile and polluted by vehicle fuels, dust from asphalt eroded by studded car tyres, wood burned for heat and smoke from ships in port. Other than that, air quality in Norway is generally good. Water quality is also generally good as far as emissions from sources on land are concerned, but there are locally polluted sea-bottom sediments from earlier times and shorelines polluted by garbage dropped from ships. On some occasions there have been larger oil spills from ships, which do not appear to have inflicted lasting damage on the natural environment.

In the second half of the twentieth century, acid rain originating from the European continent and the United Kingdom was a serious problem, causing extinctions of freshwater fish in large parts of Southern Norway. This problem has been largely solved through emission control in power plants and industries burning coal in Europe, and through mitigating measures such as injections of chalk and reintroduction of fish in affected Norwegian inland waters.

Forests in Norway have grown since the 1950s to roughly twice the volume of standing timber. This is caused by a combination of slack demand from the paper and other wood-based industries, less extensive use of lands for grazing and improved growth resulting from a warmer climate. Fauna in woodlands is generally well sustained, also because hunting is common but regulated and effectively controlled. The population of wolves in Norway is an enduring source of controversy, mainly between environmental interests calling for a sustained population of wolves and rural interests wanting to keep wolves away from their homes and their grazing livestock. In a survey in 2016–17, a total of 110 wolves were found in Norway, half of which lived in areas straddling the border with Sweden.[35] There are complaints of sheep being lost to wolves and to certain other large predators.

Fishing in Norwegian waters is generally effectively regulated and stocks monitored. Most stocks of fish tend to migrate between Norwegian waters and other waters, and effective management therefore depends on cooperation with other nations. During parts of the twentieth century, there were concerns over depletion of important fish stocks from overfishing, which remains a pervasive problem in many oceans of the world. There are still concerns for vulnerable parts of marine life and for certain seabird populations threatened by reduced access to the ocean species on which they feed.

---

[35] www.rovdata.no/Ulv/Bestandsstatus.aspx, last accessed 10 January 2018.

2.10   COOPERATION BETWEEN THE INDUSTRY AND THE GOVERNMENT

In the 1920s–30s, labour conflict caused Norway to lose more than 1 million days of work per year on average. Since the Second World War, this number has been much lower: 124 000 per year on average for the entire period from 1945–2014, corresponding to about 0.05 per cent of time worked. Most years in that period saw fewer than 100 000 days lost and in some years there were nearly none; but there were also certain years of major conflicts costing more than 500 000 days of lost work (1956, 1986, 1996 and 2010).

The lower levels of labour conflict since the Second World War reflect more settled and constructive forms of interaction between employers and employees, represented by their organizations. The government also contributes to what has become known as the 'tripartite' model of cooperation in Norwegian economic life. Norway has a main business federation organizing most enterprises which have some employees. The main labour federation has long-established links to the Labour Party. There are also some lesser labour unions not linked to the main federation or to any political party. The main labour agreements for organized personnel are negotiated every second year, with minor adjustments to be agreed in the other years. For this reason, the statistics for days of labour lost to conflict tend to show high levels in the even-numbered years, which is when the main agreements are negotiated and the potential for conflict is higher. On some occasions, pending conflicts between the employer federation and the labour federation have been resolved by government intervention, which could provide some easements on taxes, pensions or other conditions which are important to the federations.

Norway has a system for compulsory arbitration of labour conflicts. In cases where a labour conflict threatens lives or personal safety, or may have other particularly grave consequences for third parties, the government can propose fast-track legislation ordering work to be resumed and the pending matter to be resolved through arbitration. This facility has been used on several occasions and such orders to resume work are generally respected promptly. The parties in conflict often seek to avoid compulsory arbitration by avoiding stoppages which could endanger lives, safety and so on.

The practices of cooperation between government, businesses and employees in economic life evolved in the years of post-war reconstruction from

1945 (Bull 1979,[36] Hodne and Grytten 2002).[37] Effective reconstruction and economic growth were commonly agreed objectives. The Labour Party, then the governing party, had abandoned its earlier attachments to dogmatic socialism in favour of policies of a mixed economy framework with essential contributions from private enterprise, but with strong political guidance. In the early post-war years, politicians and businesses generally accepted the need for micro-managing the economy with licence requirements for imports and several other economic activities, in order to manage the scarcity of resources. Such measures were mostly abandoned in the late 1940s and the 1950s. Economic policy then aimed for strong growth in production, affluence and welfare, the means to which included selected state investments in enterprise as well as social reform. The labour federation endorsed such policies enthusiastically, while the employer federation did so with some reservations, preferring a larger role for private enterprise and more cautious progress on social reforms than that promoted by the Labour governments.

After 1970, the priorities for economic policy gradually shifted. The 1980s saw a revival of more market-oriented policies and later decades have seen more concern with maintaining the competitiveness and viability of non-petroleum and non-public sectors of the economy. Employer interests, trade federations and the State continue to cooperate on a number of issues in which they find common interests. Examples include joint efforts to reduce absences for sickness and efforts to roll back the informal economy, which is probably smaller in Norway than in many other countries in relative terms, but is perceived as a threat to common interests nevertheless.[38]

2.11 CONCLUSION

Al-Kasim (2006) has pointed out that Norwegian society was in a good position to embark on petroleum developments as these started around 1970 – much more so than many countries which have embarked on such developments more recently.[39] Norway as of 1970 was a rather well-organized society, with firmly established democratic traditions and legal system, a fairly sophisticated economy even by 2018 global standards, an effective educational

---

[36]   Edvard Bull, '*Norge i den rike verden*' in *Norges historie*, vol 14 (J.W. Cappelens Forlag 1979).

[37]   Fritz Hodne and Ola Grytten, *Norsk økonomi i det 20. århundre* (Fagbokforlaget 2002).

[38]   www.nho.no/Politikk-og-analyse/Seriost-arbeidsliv-og-samfunnsansvar/NHO -om-svart-arbeid/, last accessed 10 January 2018.

[39]   Farouk Al-Kasim, *Managing Petroleum Resources: The 'Norwegian Model' in a Broad Perspective* (Oxford Institute of Energy Studies, 2006).

system and a competent workforce with capabilities relevant to petroleum developments. After some initial dithering, the country's leadership had the foresight and patience to proceed on petroleum developments with caution, confident that the country was in no urgent need and that neither the oil in the ground nor the demand for it were likely to go away any time soon.

Since 1970, petroleum revenues and sector activities have helped Norway to proceed further in the directions already laid out: increased affluence, an even more resourceful State and more sophisticated legislation and frameworks to manage the important aspects of society.

Other countries can emulate these conditions to some extent, but certainly not to the full extent. Another factor not to be underestimated is Norway's sheer luck with its petroleum. The resources were very valuable. The oil price increases in 1974 and 1980 came at times which were clearly fortunate for Norway. There was a good match of geological conditions, technological developments, Norwegian established capabilities and the timing of business cycles in other industries making such capabilities available on the labour market. Norway could apply a degree of soft protectionism favouring its own industries as participants in and suppliers to the petroleum sector, just before new international accords impeded such practices. The relevance of lessons which other countries may learn from Norway is moderated by inevitable differences in circumstance and by the fact that luck – not only the quality of decisions made – also contributed to the fortunate outcomes.

3. Hydrocarbon policy and legislation: Norway

Tonje Pareli Gormley and Merete Kristensen

3.1   INTRODUCTION

Norwegian petroleum policy and the legal framework for petroleum activities are characterized by strong host country control. The State is the owner of petroleum resources and has the exclusive right to resource management.[1] It carries out resource management based on policies as implemented in legislation and through the exercise of administrative authority.[2] The licensing regime is the primary vehicle for Norwegian host country control of resource management, in combination with requirements for additional permits, approvals and consents, and the power to stipulate conditions for issuing the same.[3] This is not an unusual model for resource management in modern international petroleum law. However, the use of a licensing model as a tool for natural resource management in Norway dates back to the 1909 regulation of hydropower. By the time the petroleum era commenced, the Norwegian State already had longstanding experience in implementing such legislation.[4]

---

[1]   The Petroleum Activities Act of 29 November 1996 Number 72 (the '1996 Act'), Section 1–1. The allocation of powers for resource management is further detailed in Section 1–2.

[2]   Norwegian petroleum law is intertwined with, *inter alia*, Norwegian administrative law. This chapter focuses on the sector-specific legal framework for petroleum activities and not on administrative law.

[3]   The 1996 Act, Section 10–18, third paragraph, reads: 'In connection with individual administrative decisions, other conditions than those mentioned in this Act may be stipulated, when they are naturally linked with the measures or the activities to which the individual administrative decision relates.'

[4]   As stated in *Storting* White Paper 28 (2010–11), item 1.1: 'The legislation stipulated the right of reversion (to the State), emphasised that the Norwegian people are the owners of the water resources, and that economic rent should fall to the greater community'. These same principles have been followed in the administration of the petroleum resources. www.regjeringen.no/globalassets/upload/oed/petroleumsmeldingen_2011/oversettelse/chapter1_white_paper_28-2010-2011.pdf, last accessed 1 May 2018.

Moreover, as Norway was an established industrialized nation, the Norwegian authorities were already experienced at developing industry policies to encourage development. Therefore, we would argue that it is the manner in which the Norwegian model has been implemented and developed over the decades that has proved to be its strength.

Norwegian petroleum policy, petroleum legislation and state organization of the industry have developed concurrently with the Norwegian petroleum industry. This development has been shaped by the knowledge and experience gained and the need to respond to contextual changes over time.[5]

In the late 1950s, few could have predicted the tremendous potential of Norway's resource base. In 1958 the Norwegian Geological Survey concluded: 'The chances of finding coal, oil or sulphur on the continental shelf off the Norwegian coast can be discounted.'[6] However, discoveries of petroleum were made offshore the Netherlands just the following year and international oil companies started to take an interest in the North Sea. In 1962 Phillips applied for concessions over the entire Norwegian Continental Shelf (NCS). The application was rejected; it was simply asking for too much, too soon. There was acknowledgement that if petroleum activities were to be allowed on the NCS, there would be a need for competition and diversification.[7] Moreover, there was consensus that at the time, it was still too early to commence petroleum activities on the NCS. The Norwegian authorities did not yet have the requisite knowledge of the petroleum industry and there was no legislation for petroleum activities in place. However, as a result of the Phillips application, Norway recognized the need to define petroleum policies, to implement these in legislation and to build institutional competence and capacity.

---

[5]   Examples of contextual changes include the perceived resource base of a host country, the host government's policy goals, international public law obligations, other specific activities and events (eg, technological development, market conditions such as oil prices and cost levels), and the capacity and competence of the host government's institutions.

[6]   www.npd.no/en/Publications/Norwegian-Continental-Shelf/No1-2006/ Fragments-from-another-Norwegian-oil-history/, last accessed 16 November 2017.

[7]   As the present-day Norwegian petroleum authorities have observed: 'The company asked for a license for the parts of the North Sea that were in Norwegian territorial waters and that were or might be designated as part of the Norwegian continental shelf, and offered USD 160 000 per month. This was regarded as an attempt by the company to obtain exclusive rights. The authorities decided that it was out of the question to hand over the entire continental shelf to one company. If these areas were to be opened for exploration, more companies would need to be involved'. www .norskpetroleum.no/en/framework/norways-petroleum-history, last accessed 3 August 2016.

The rejection of the Phillips application reflects the Norwegian authorities' knowledge and experience-based policy for resource management and regulatory development. Although the authorities were quick to establish a basic legal framework so that petroleum activities could commence shortly after the Phillips application was rejected, there was no rush to establish a comprehensive legislative framework for petroleum activities. While there was agreement that regulation was required, emphasis was put on the need to learn more about the industry and gather experience before adopting a comprehensive law.[8] The overarching approach to regulatory development was to adopt the most essential rules first, and then develop more detailed rules as knowledge and experience were gained.

The overall objectives of Norway's current petroleum policy are as follows:

> to provide a framework for the profitable production of oil and gas in the long term. The value creation shall benefit the Norwegian society as a whole, activities must take place within a sound HSE framework, and environmental concerns and coexistence with other industries are to be taken into account throughout the industry.[9]

Although this quote relates to present-day policy, more than 50 years after the first Norwegian licensing round, it still reflects the core of Norwegian petroleum policy over time. The priority has been to create and, insofar as is reasonable, maintain a stable and predictable framework for petroleum activities. However, as former Minister of Oil and Energy Tord Lien stated in Mexico City in 2014: 'Stability is not the same as static. Policies and measures have changed due to new realities on the Shelf or in the global petroleum industry. Adapting to changing realities is very important to create maximum value.'[10]

This quote acknowledges the need for the host country to balance, adequately and over time, the interests of the State as host country with those of the companies operating within the petroleum sector. The petroleum industry is capital intensive, high risk and subject to rapid technological development. By maintaining close dialogue with the industry and demonstrating a certain degree of responsiveness to the changing needs of industry players, Norway has remained relatively attractive for investments over time.[11] Indeed, this

---

[8] NOU 1979:43, Chapter 1.

[9] Quote from www.petroleum.no, an information website run by the Ministry of Petroleum and Energy and the Norwegian Petroleum Directorate. The quote is the opening statement on the information given on the framework; www.norskpetroleum .no/en/framework/, last accessed 3 November 2017.

[10] www.regjeringen.no/no/aktuelt/Norways-petroleum-activities-and-related -industry-development/id2008375/, last accessed 1 May 2018.

[11] For a more detailed analysis on the interaction between the Norwegian authorities and licensees, see *inter alia* Farouk Al-Kasim, *Managing Petroleum Resources –*

approach to resource management is still deemed necessary to achieve Norway's main policy goals:

> The main elements of the petroleum policy are in place. It is important to build further on the successful resource management. The main challenge to achieve the goal of the policy is increased recovery from fields, development of discoveries and to establish undiscovered resources. To achieve this it is important that adjustments are made in the use of policy instruments when the development in the industry and/ or the resource base indicates that this is appropriate. The interaction between state, oil companies, the supply industry and the research sector is an important part of Norwegian resource management.[12]

Today, Norway is one of the world's major producers. Although parts of the NCS are considered mature, new thinking has nevertheless resulted in major discoveries in mature areas in recent years. In 2007 the Edvard Grieg field was discovered and in 2010 the Johan Sverdrup field was discovered. In 2016 the first exploration activities commenced in the Barents Sea. Industry interest in these newly opened areas has been considerable. There is therefore reason to believe that the Norwegian petroleum industry will survive for a good few decades yet. In any event, because of the successful implementation of Norwegian local content policies, among other things, Norwegian oil companies and the Norwegian oil service industry have proved to be competitive on an international scale.[13] Thus, when Norway eventually runs out of petroleum, Norwegian players will be able to continue operations globally.

Norway's political and legislative history in relation to petroleum activities and resource management is extensive. This chapter focuses on those features of Norwegian policy and legislation which may help to explain the success of the Norwegian model, from which other host countries may wish to take inspiration in developing their own domestic policy and legislation.[14] However, those seeking inspiration from the Norwegian experience should take note that Norwegian petroleum policy and the manner in which it is implemented are tailored to the Norwegian context – both factually and legally. Whereas the Norwegian experience offers many learning points, domestic petroleum policy and legislation must always be tailored to suit the domestic context.

The 'Norwegian Model' in a Broad Perspective, Oxford Institute for Energy Studies, 2006 ('Al-Kasim'), Chapter 8, 'Value creation: A common objective', pp149 et seq.

[12]   As stated in white paper *St meld* nr 28 (2010–11), p6. The quote is an unofficial office translation.

[13]   See item 1.5 on local content.

[14]   Due to the scope of this chapter, we will not give a full account of every development in the Norwegian legal regime, but rather highlight milestones which may shed light on important developments during the past 50 years.

3.2 HISTORICAL EVOLUTION OF NORWEGIAN HYDROCARBON POLICY AND LEGISLATION[15]

3.2.1 The Core Principles of Norwegian Petroleum Policy

Despite the rapid development of the Norwegian petroleum industry and several changes in government, the main policy goal for the Norwegian petroleum industry has always been to provide a framework for the profitable production of oil and gas in the long term.[16] Moreover:

> It has also been considered important to ensure that as large as possible share of the value creation accrues to the state, so that it can benefit society. Oil and gas activities must take place within a sound health, safety and working environment framework, and environmental concerns are to be taken into account throughout the industry. Petroleum activities must also take place in coexistence with other industries.[17]

The above can be characterized as the core of Norwegian petroleum policy as expressed in a string of reports, or white papers, from government to Parliament over the past five decades.[18]

At the dawn of the Norwegian petroleum era, some of the most important policy goals addressed fundamental issues such as securing sovereignty over natural resources and the exclusive right to resource management, and promoting local content and state participation.

In 1971 a set of guiding principles on resource management was established in a report to Parliament.[19] These principles are popularly referred to as the 'Ten Oil Commandments', as they have greatly influenced the direction of

---

[15]   This chapter is written with a focus on the development of sector-specific petroleum legislation, such as acts of Parliament concerning petroleum activities and regulations adopted thereunder. Although an important part of the overall Norwegian approach to resource management, the chapter does not discuss the development of other interfacing legislation such as laws on public administration, freedom of information, environment and working environment.

[16]   This is expressed in a number of sources, including recently in the government's February 2017 Strategy for the Sea, www.regjeringen.no/no/dokumenter/ny-vekst-stolt -historie/id2552578/, p41, last accessed 1 November 2017.

[17]   See www.norskpetroleum.no/en/framework/fundamental-regulatory-principles/, last accessed 24 April 2017.

[18]   It is an established practice that the government submits reports, or white papers, to the Parliament (in Norwegian, '*Stortingsmelding*' or abbreviated to '*St meld*') when it wishes to present matters for the Parliament without a proposal for a decision – for example, to report on work that has been carried out or to discuss policy. These reports, and the debate thereon in Parliament, often form the basis for subsequent bills.

[19]   *St meld* nr 76 (1970–71).

the Norwegian petroleum industry. These principles established, *inter alia*, that national management and control of petroleum resources in a coordinated manner is important to ensure that resource management benefits Norwegian society as a whole through security of supply and value creation in the form of business development. Moreover, the importance of prudence in handling interfaces with other interests such as the environment, existing industrial activities, socio-political issues and foreign affairs was addressed. In another report to Parliament issued two years later, the government took this one step further, establishing that petroleum resources should be used to develop a 'qualitatively better society'.[20] The importance of the development of Norwegian petroleum competency, in both the administrative and commercial sense, was also highlighted.[21]

BOX 3.1 THE TEN OIL COMMANDMENTS

1. That national supervision and control of all activity on the Norwegian Continental Shelf must be ensured.
2. That the petroleum discoveries must be exploited in a manner designed to ensure maximum independence for Norway in terms of reliance on others for supply of crude oil.
3. That new business activity must be developed, based on petroleum.
4. That the development of an oil industry must take place with necessary consideration for existing commercial activity, as well as protection of nature and the environment.
5. That flaring of exploitable gas on the Norwegian Continental Shelf must only be allowed in limited test periods.
6. That petroleum from the Norwegian Continental Shelf must, as a main rule, be landed in Norway, with the exception of special cases in which socio-political considerations warrant a different solution.
7. That the State involves itself at all reasonable levels, contributes to coordinating Norwegian interests within the Norwegian petroleum industry, and to developing an integrated Norwegian oil community with both national and international objectives.
8. That a state-owned oil company be established to safeguard the State's commercial interests, and to pursue expedient cooperation with domestic and foreign oil stakeholders.

---

[20] *St meld* nr 25 (1973–74), p6.

[21] Furthermore, Norwegian petroleum authorities have always sought to ensure that resource management is based on knowledge and facts. Great importance has also been placed on the interaction between the State and key industry stakeholders, and on the need for competition and diversification.

9. That an activity plan must be adopted for the area north of the 62nd parallel which satisfies the unique socio-political factors associated with that part of the country.
10. That Norwegian petroleum discoveries could present new tasks to Norway's foreign policy.

Source: *St meld* nr 6 (2010–11), p8.

Subsequent petroleum policy was built on the principles expressed in the Ten Oil Commandments, while being more specifically targeted at tackling new challenges and developments as they arise. We will take a closer look at how some of these targeted policies have been implemented in the Norwegian petroleum legislation in section 3.2.2.

An overall observation when studying the Norwegian experience is that Norwegian petroleum policy has largely been successfully implemented in legislation. Whereas the core principles of Norwegian petroleum policy have remained stable, their implementation has evolved over time in response to changing circumstances. This combination of relatively stable policy and a dynamic approach to implementation has been enabled, *inter alia*, by an active and open dialogue with the industry, allowing Norway to remain attractive for investors over time.

Implementation of, and compliance with, policy require strong and competent state institutions. The Norwegian State has focused on building institutions that are competent and capable of supervising and enforcing the legal framework. This has enabled a consistently high level of host government control across all phases of petroleum activities, which from a global point of view is relatively rare. Maintaining this level of host government control without compromising the investment climate can be a challenge. It requires a legal framework that ensures clarity in roles and responsibility, transparency and accountability. Moreover, an environment of low political risk is an advantage.

3.2.2 Main Milestones: Legislative Developments

Step by step, the Norwegian legal framework for petroleum activities has evolved from a very brief framework-enabling law in 1963 to the comprehensive 1996 Act and a plethora of regulations. Moreover, over time, model production licences and a Standard Joint Operating Agreement (JOA) have been developed.[22] During all of these steps, the policy of sovereignty over natural resources and strong host country control has been continuously maintained

---

[22] See www.regjeringen.no/en/find-document/dep/OED/Laws-and-rules-2/Rules/konsesjonsverk/id748087, last accessed 24 April 2019.

and implemented in legislation – among other things, through the licensing regime and the appurtenant requirements for approvals, consents and permits at various stages of petroleum activities, as well as the power to stipulate conditions for such decisions. The grant of approvals, consents and permits, as well as the power to stipulate conditions, have typically been subject to the exercise of a certain degree of discretionary powers by the petroleum authorities.[23] This model promotes a stable legal framework for petroleum activities, while presenting the opportunity to set conditions that are tailored to each specific project.

The first step towards the regulation of petroleum activities aimed to secure sovereignty over petroleum resources in the areas that today constitute the NCS. In 1963, shortly after the Phillips application, a Royal Decree[24] establishing sovereign rights to explore and exploit these natural resources was adopted. The wording was carefully phrased, as the international law of the sea was still evolving and Norway's maritime boundaries with the UK and Denmark were yet to be agreed. Thus, on one hand, the Decree made specific reference to the median line principle; while on the other hand, it avoided the use of the term 'continental shelf':

> The sea-bed and the subsoil in the submarine areas outside the coast of the Kingdom of Norway are under Norwegian sovereignty in respect of the exploitation and exploration of natural deposits, to such extent as the depth of the sea permits the utilisation of natural deposits, irrespective of any other territorial limits at sea, but not beyond the median line in relation to other states.[25]

Later that year, a brief framework law on subsea natural resources was adopted (the '1963 Act').[26] The 1963 Act merely established the State's exclusive right to subsea natural resources, the right for the King[27] to grant Norwegian and foreign persons and companies licences to explore for and exploit subsea natural resources, and authorization for the King to stipulate conditions for the grant of such licences and issue rules on such activities.

---

[23]   Whereas broad discretionary powers are viewed as an advantage from a host government's point of view, they can be viewed as a risk from an investor's point of view.

[24]   Royal Decree of 31 May 1963. Norway acceded to the Geneva Convention on the Continental Shelf of 29 April 1958 on 9 September 1971.

[25]   Unofficial English translation of the 1963 Decree, included in *Lovgvning vedrørende den norske kontinentalsokkel*, Royal Ministry of Industry and Handicrafts, Oslo, January 1973.

[26]   Act 12 of 21 June 1963 on exploration and exploitation of subsea petroleum resources.

[27]   That is, the government.

In 1965 a Royal Decree[28] was adopted that implemented the 1963 Act (the '1965 Decree'). The 1965 Decree established the main rules on petroleum activities, safety and the environment. Importantly, the 1965 Decree introduced the Norwegian licensing regime. The main contours of the licensing regime established under the 1965 Decree, which differentiated between reconnaissance licences and production licences, are still reflected in the current licensing regime. Under Section 14 of the 1965 Decree, licensees were granted the exclusive right to exploration and exploitation of petroleum within the relevant licence area. Licensees were also, as a starting point, granted the right to take and dispose of petroleum produced. However, this right was subject to exceptions. Among these, the indirect power of the Norwegian authorities to require petroleum to be taken ashore in Norway through the exercise of their powers under Section 35 to approve, among other things, the location of certain facilities for transportation and storage was subject to much debate.[29] The Norwegian authorities acknowledged, however, that this power should be exercised with care. In 1968 a committee evaluated the possibility to bring gas and condensate ashore from the North Sea to Norway. The committee concluded that a requirement that gas and condensate be brought ashore to Norway could not yet be recommended, given the geographical challenges posed by the Norwegian trench ('*Norskerenna*')[30] and the lack of an adequate gas market in Norway. The committee stated that a different conclusion might be reached in the case of oil discoveries. The issue of allowing licensees to take petroleum ashore in other countries was of great political importance. It was noted in a 1969 report to the Parliament that whereas licensees had the right to petroleum produced, petroleum was nevertheless a Norwegian natural resource that should be utilized in such a manner as was most advantageous for Norway. Thus, how the power to require petroleum to be brought ashore to Norway should be exercised was to be evaluated specifically in each case, with a view to finding solutions that were the most advantageous for Norway while taking the interests of licensees into consideration. It was deemed too early to draw up general guidelines on how this power should be exercised. This rather non-committal position represented an element of insecurity for licensees. The

---

[28] Royal Decree of 9 April 1965.

[29] In addition, Article 33 established that the King (ie, the government) had the power to require that petroleum was to be brought, in whole or in part, ashore to Norway when 'national interests so dictated'. What constituted 'national interests' was down to the discretion of the King; however, the licensee was given the right to provide its comments before any decision was made. Under Article 34, the King also had the right to require that petroleum produced be placed at the disposal of the Norwegian authorities in case of war or national crisis.

[30] A subsea trench with a depth of between 200–700 metres.

authorities acknowledged this, stating in response that it would be reasonable to ensure that decisions on this matter would be made efficiently.[31] The 1965 Decree also established rules on the regular submission of data, samples and information to the Ministry of Industry, on the relinquishment of areas and on the assignment of participating interests, which were to be approved in writing by the Ministry.

It was only after the adoption of this set of main rules in the 1965 Decree that the first licences were granted, in September 1965. This first licensing round was also the most generous licensing round that Norway has carried out: 278 out of 314 blocks were open for bids. This was due to the desire to explore as much as possible of the NCS to gather data, which in turn would serve as a basis for making informed decisions on future licensing policy. Licences were granted to nine companies, which undertook to carry out seismic work and drill a total of 30 wells within a six-year period.[32]

The first major discovery, the Ekofisk field, was made in 1969. Production commenced in 1971. Other major discoveries were made in the years thereafter.[33] As Norwegian petroleum activities expanded in both scope and volume, the legislative framework was continually developed and adjusted.[34] By the early 1970s, Norwegian competence and capacity in petroleum-related issues had been considerably strengthened. Moreover, several discoveries had made the NCS more attractive. Thus, there was room for Norway to tighten the conditions for petroleum activities. The Ministry stated in a 1970 report to the Parliament that the time had come to establish more comprehensive regulation of petroleum activities. However, it was expected that developing a comprehensive petroleum law would be time consuming. Therefore, as a temporary solution, the 1965 Decree was replaced with a more comprehensive version in 1972 (the '1972 Decree').[35] The 1972 Decree was built on the same main principles as the 1965 Decree, but the licence conditions were generally stricter. Among other things, Section 34 stated that petroleum produced was required to be brought ashore to Norway unless the Ministry, upon application, approved otherwise.

---

[31]   Report 95 to the Parliament (1969–70), pp19 et seq.

[32]   Report 95 to the Parliament (1969–70), p1.

[33]   For a more detailed overview, see www.norskpetroleum.no/en/framework/norways-petroleum-history, last accessed 4 August 2016.

[34]   The first law on petroleum tax was adopted in 1965 (Act Number 3 of 11 June 1965). In addition, other regulations were gradually developed and issued under the 1963 Act – that is, the 1967 safety regulations on drilling (updated and replaced in 1975), the 1976 regulations on production and the 1978 conservation regulations.

[35]   Report 76 to the Parliament (1970–71), p16 and Royal Decree of 8 December 1972.

It was only in 1985 that the comprehensive new petroleum law and appurtenant regulations were adopted (the '1985 Act').[36] This represented a major step forward in Norway's legislative development. The law, which took considerable time and effort to draft, was built on both past domestic experience and broad international experience. Section 3 of the 1985 Act explicitly established that resource management was to be carried out so as to ensure that consideration was given to other activities and the environment, to the benefit of Norwegian society as a whole.[37]Although phrased as a policy statement, this would have an impact on the exercise of discretionary powers under the 1985 Act and regulations.

The 1985 Act covered all phases of a petroleum project, from reconnaissance to abandonment and decommissioning. It established a number of new rules and rules that codified existing practices. For instance, rules were established on the opening of new areas for petroleum activities, and rules on development and production were codified and further developed. The 1985 Act also established rules on a petroleum registry and on mortgages.

Importantly, the 1985 Act clarified the roles of the various petroleum authorities. The Norwegian Petroleum Directorate (NPD) was assigned a coordination role, as part of which it was charged with preparing a coordinated and general regulatory regime for health, safety and the environment (HSE). The period immediately after the 1985 Act was adopted was characterized by an effort to tidy up what was now a fragmented regulatory framework on HSE in the petroleum industry. As a first step, the existing regulations were simplified and aligned with the principles of the 1985 Act. There was a gradual shift away from detailed, prescriptive rules towards more functional, goal-based rules. In 1992, 13 new regulations addressing different topics were adopted. As from 1993, the Working Environment Act expanded its scope to cover the petroleum industry. The work on establishing a new regulatory framework for HSE in the petroleum industry led to the adoption of a regulation for systematic follow-up on the working environment in the petroleum industry in 1995.

Despite the lengthy legislative process and all the work put into its adoption, the 1985 Act was still amended from time to time to adapt to changing circumstances and new challenges. Subsequent amendments included, for instance, the introduction of a new chapter on compensation for fishermen (1989) and adaption to the European Economic Area (EEA) Agreement,[38] including

---

[36]   Act 3 of 22 November 1985.
[37]   In Section 3.
[38]   Available in English at www.efta.int/legal-texts/eea, last accessed 24 April 2019. The EEA Agreement effectively requires Norway to comply with certain parts of EU regulations. These include requirements for the free movement of goods, services, capital and persons. The EEA Agreement was entered into between the then European

removal of local content requirements and implementation of the Licensing Directive (94/22/EC).

After only a few years in force, therefore, it became clear that there was a need for a comprehensive revision of the 1985 Act. On 29 November 1996, the 1996 Act was adopted. The stated reasons for the revision[39] included a need to adjust the law to reflect the new phase that the industry was moving into, with a continued emphasis on rational and efficient solutions while ensuring a high level of HSE protection across all activities. There was also a need to reorganize its chapters and provisions.

The 1996 Act established in Section 1–4 that it applies to petroleum activities in connection with subsea petroleum deposits under the Norwegian jurisdiction. This means that the 1996 Act applies to such activities whether they are carried out onshore or offshore. Compared to the 1985 Act, this represented a clarification in scope. The rules on the duration of production licences were also amended. Among other things, the longest possible extension period of a production licence was amended from 30 to 50 years (Section 3–9). A new provision on dividing the area of a production licence was also introduced as an alternative for the unitization and coordination of petroleum activities (Section 3–10); as were rules to simplify the administration relating to development plans for small discoveries (Section 4–2, sixth paragraph). In Chapter 5, new and more detailed rules on abandonment and decommissioning were established. The 1996 Act further introduced a requirement for licensees to submit a decommissioning plan two to five years prior to the cessation of use of a facility or the expiry of a licence. Moreover, a new provision was introduced to codify the established practice that petroleum activities shall be conducted in such a manner as enables a high level of safety to be maintained and further developed as accords with technological developments (Section 9–1). The scope of the obligation to provide material and information regarding petroleum activities to the authorities was extended to encompass any person conducting petroleum activities – not merely licensees (Section 10–4).

The 1996 Act is comprehensive and a plethora of regulations have been issued thereunder. Of particular importance in the development of regulations was the continuous work to create a coordinated HSE regulatory framework. Thus, in 2001 the number of HSE regulations was reduced to five general HSE regulations issued jointly by the NPD, the Health Directorate and the then Norwegian Pollution Control Authority. The functional, goal-based require-

---

Community and the European Community's member states on the one hand and the European Free Trade Association States – Iceland, the Principality of Liechtenstein and the Kingdom of Norway – on the other.

[39] *Ot prp* nr 43 (1995–96), item 1.

ments were supported by non-binding guidelines. This approach to HSE regulation was unique globally.[40]

In 2001 the 1996 Act was once again amended to include a new Chapter 11 on the State's direct financial interest. This amendment, which was prompted by the partial privatization of national oil company Statoil (now Equinor), is discussed in further detail in section 3.5.

In 2003 the scope of the 1996 Act was expanded to clearly state that the Act applies to petroleum activities relating to subsea deposits, including, under certain conditions, onshore activities relating to the utilization of produced petroleum that takes place on Norwegian land or seabed territory, subject to private property rights only, when such utilization is necessary to or constitutes an integrated part of production or transportation of petroleum (Section 1–4, second paragraph). Consequently, the HSE regime was amended yet again in 2011 to include a general regulatory framework for petroleum activities both onshore and offshore. In 2003 a new Section 4–9 on extended operator responsibility for the overall operation of upstream pipeline network was also introduced. This amendment was triggered by the establishment of Gassled, with Gassco as operator, and is further discussed in section 3.5. Section 4–8 on third-party access was also amended with respect to the stipulation of tariffs, other conditions and the amendment thereof.

The next wave of revisions came in 2009. These included, among other things, the introduction of a new rule in Section 5–3 on an assignor's alternative liability for financial obligations towards the remaining licensees for the cost of carrying out decisions relating to abandonment and decommissioning. The scope of Section 4–8 on third-party access was expanded to include facilities that are used, but not owned by a licensee, and that may be used for the treatment, transportation and storage of carbon dioxide.

3.2.3 Summary of Observations

The above overview of the main legislative milestones shows how the Norwegian authorities have kept a close eye on the petroleum industry and any changes in framework conditions, whether in respect of the market, technological developments or new policy objectives spurred by the increasing maturity of the NCS.

The development of the Norwegian regulatory framework for petroleum activities also shows how Norway has had a continuous focus on strict host

---

[40] See www.ptil.no/sss2016e/the-foundation-article11860-1222.html, last accessed 10 May 2018.

country management and control. This is also apparent in the following sections on state participation and local content development.

Today, the comprehensive 1996 Act, the regulations issued thereunder, the production licence and the Standard JOA[41] constitute the main instruments through which the Norwegian petroleum authorities manage the country's petroleum resources. While these instruments have established a stable, predictable framework, they also leave a considerable degree of discretionary power to the authorities. Despite the overall high level of continuity and predictability in policy, and the safeguards in administrative law, one may question whether the manner in which the Norwegian authorities have exercised these powers is sufficiently transparent. Moreover, as the powers of the Ministry of Petroleum and Energy (MPE) are so wide, one may also question whether the systems for accountability as established are working. Companies may be reluctant to lodge complaints against such a powerful body. For instance, we are not aware of any complaints being lodged by companies that have not been awarded licences in licensing rounds. We note, however, that recent history may indicate an increased level of lawsuits challenging the exercise of discretionary powers, including the *Gassled* case.

3.3   NORWEGIAN STATE ORGANIZATION OF UPSTREAM ACTIVITIES

At the dawn of the petroleum era on the NCS, Norway already had a well-functioning state administration which was based on the separation of powers between the legislative branch (Parliament), the executive branch (the government, supported by the state administration), and the judicial branch (the courts). In addition to its legislative powers, the parliamentary system was well established and Parliament already exercised oversight over the executive branch. The courts were also viewed as independent and unbiased. The political situation was relatively stable. The state administration was characterized by able, stable institutions that promoted sound business practices to encourage industrial dynamics and help shape industrial policies. These institutions operated on the basis of prudent administrative practices, characterized by a relatively high level of predictability, transparency and accountability. The Norwegian Public Administration Act[42] was enacted in 1967. Aiming, among other things, to safeguard the rule of law, thoroughness and impartiality in public administration, it established general rules on legal competence, case

---

[41]   See   www.regjeringen.no/en/find-document/dep/OED/Laws-and-rules-2/Rules/konsesjonsverk/id748087/, last accessed 24 April 2019.
[42]   Public Administration Act of 10 February 1967.

handling, case preparation prior to making decisions, complaints and reversals and regulations.[43] The Freedom of Information Act was enacted in 1970. Subject to defined exceptions, this act established the principle of free access to information in public documents for the public in general.[44] This was certainly an advantageous starting point.

At first, as the natural choice, the Ministry of Industry was charged with the management of the petroleum industry. Norway wished to maintain tight state control of petroleum activities on the NCS.[45] There was thus a need for increased specialized competence and capacity, in addition to the institutions that already existed. In 1963 a continental shelf panel was appointed to assist the Ministry of Industry with the preparation of laws and regulations. In addition, the State Petroleum Committee[46] was established in 1965 in order to assist the Ministry of Industry with matters relating to petroleum exploration and exploitation. In 1966 a separate petroleum section was established within the Ministry of Industry.

Despite the need for specialized competence, there was also a need to regulate the interface with existing institutions. In 1967 regulations on safety in the petroleum sector were issued by the Ministry of Industry.[47] As the regulations affected the areas of responsibility of a number of other state entities, the power to carry out inspections was delegated from the Ministry of Industry to a number of other authorities, including the Maritime Directorate, the Labour Inspection Authority and the Directorate of Health.[48] Various sector-specific authorities handled environmental issues relating to their respective areas of responsibility until the Ministry of Environment was established in 1972, as the world's first ministry with responsibility for overall environmental management.

Nevertheless, as the potential of the Norwegian petroleum sector for the Norwegian economy and society as a whole became clear, so too did an under-

---

[43]   By and large, this was a codification of previous practices for administrative case handling. Norwegian administrative law is built on certain key principles, such as the principle of legality, as well as principles aimed at ensuring that any discretionary decisions are properly made (eg, the principle of proportionality).

[44]   Freedom of Information Act of 19 June 1970.

[45]   Since the commencement of petroleum activities, Norway has had a licence-based regime for granting exploration and production rights. Licences are typically granted on certain conditions, with which licensees must comply. Moreover, in order to move forward with petroleum activities or a specific project, governmental approvals and consents are typically required.

[46]   In Norwegian, *Statens Oljeråd*.

[47]   Royal Decree of 25 August 1967.

[48]   Report 95 to Parliament 1969–70, p7.

standing that the capacity and competence of the authorities should follow suit. Al-Kasim comments:

> In a report to the Storting submitted on 12 June 1970, the Ministry was cautious regarding the future organisation of its functions within the petroleum sector. It stated amongst other things that 'the Petroleum section has been established, and the personnel have been engaged, on a provisional basis. This form of organisation has been adopted because the Ministry does not yet know whether there will be permanent activities on the Continental Shelf. . . In the longer run, the present organisation form will be undesirable. It may be necessary to establish a directorate for Continental Shelf matters, possible also a state-owned company in charge of the Government's business interests in petroleum exploration, of commercial finds of petroleum are made in one of more of the blocks where the Government has participation rights.'[49]

The perception of petroleum potential and the sustainability of a Norwegian petroleum industry on the NCS soon changed with the Ekofisk field coming on stream in 1971. Thus, by June 1972 a new structure had been adopted for the state administration of the petroleum industry. This new regime was based on the recommendations of a special committee charged by the government with recommending a new organizational model for the administration of petroleum matters. The committee had, *inter alia*, studied the examples of other, more mature petroleum countries. It identified three main functions of the host country: the centralized control function, the administrative function and the commercial function. As regards the centralized control function, it was recognized that, apart from what lies constitutionally with the Parliament and the government, day-to-day matters naturally belong with a sectorial Ministry. The Ministry of Industry, strengthened by a new Petroleum and Mining Division, was therefore made responsible for policy making, legislation and licencing. The newly established NPD, subordinate to the Ministry of Industry, was made responsible for the administrative function, including technical control and regulatory and advisory functions. Finally, Den norske stats oljeselskap (later Statoil and today Equinor ASA) was established as a wholly state-owned company, charged with commercial participation.

In 1978 the MPE was established and petroleum-related responsibilities were transferred from the Ministry of Industry to the MPE. In 1979 those responsibilities relating to safety, working environment and emergency preparedness in the petroleum industry were transferred from the MPE to the Ministry of Labour and Local Affairs.[50] The MPE was thus superior to the NPD with regard to functions relating to resource management, while the

---

49   Al-Kasim, p46.
50   In 2001 these responsibilities were transferred to the Ministry of Labour.

Ministry of Labour and Local Affairs was superior to the NPD with regard to functions relating to HSE. The rationale for this managerial split was the risk of a potential conflict of interests between achieving resource management and licensing policy goals on the one hand and maintaining a high level of HSE protection on the other.

Moreover, after the accident on the Alexander Kielland rig in 1980, when 123 persons lost their lives, the importance of coordination and information sharing between the various authorities with independent responsibilities was highlighted. Thus, in 1985 the NPD was assigned a coordination role, and many of the authorities that previously had independent supervisory powers were rather tasked with assisting the NPD. Other authorities maintained their independent role; however, the NPD was made responsible for coordination, in order to streamline their external representation, including their respective regulatory frameworks. This arrangement was welcomed by licensees, as it eased compliance.

Later, in 2004, the NPD itself was split into two bodies: the NPD and the Petroleum Safety Authority (PSA). As a result, today the NPD is responsible for resource management issues and the PSA is responsible for HSE issues.[51] The PSA is also the coordinating authority in relation to other authorities with independent responsibilities relating to health and safety, including the Norwegian Environment Agency, the Health Directorate and the Labour Inspection Authority.[52]

Today, Article 1–2 (1) of the 1996 Act reads: 'Resource management is executed by the King in accordance with the provisions of this Act and decisions made by the *Storting* (Parliament).'

The main state institutions involved in petroleum resource management and HSE in the petroleum industry include the MPE, the Ministry of Labour and Social Affairs, the NPD and the PSA. As regards environmental issues, the main state institutions include the Ministry of Environment and the Environment Agency. Coordination is still crucial. Parliament exercises parliamentary oversight over the various entities in the executive branch and is kept informed on the development of the petroleum sector through regular comprehensive reports from the executive branch on the general status of the sector, as well as shorter notifications on key decisions contemplated (eg, in relation to development plans).

---

[51] See www.ptil.no/role-and-area-of-responsibility/category916.html, last accessed 10 May 2018.

[52] See Royal Decree of 19 December 2003 on, *inter alia*, the establishment of the PSA.

In conclusion, a review of the development of the Norwegian state organization as regards the petroleum industry shows that great emphasis was initially put on capacity building to create competent institutions. Thereafter, the primary focus shifted to ensuring clarity of roles and responsibilities. This is also apparent in the account on state participation in section 3.5. These elements, combined with transparency and accountability, form part of the internationally recommended principles for good governance in the petroleum industry. Thus, it is evident that Norway's focus on establishing and implementing a system of good governance has been a contributing factor to the country's success in petroleum resource management.

3.4   NORWEGIAN STATE PARTICIPATION IN UPSTREAM ACTIVITIES

3.4.1   Introduction

A dynamic approach to state participation in upstream activities has been a key element of Norway's success in petroleum resource management. It has ensured increased government take and helped to build domestic competence within, and enhanced know-how about, the petroleum sector, thus enabling Norway to build a long-term industry that benefits society as a whole.

Over the past five decades, the Norwegian approach to state participation in upstream petroleum activities has undergone many different phases: from no direct state participation at all in the first licensing round to negotiations with prospective licensees on state participation in each licence. Thereafter, a regime of preferential rights to facilitate state participation through the national oil company was established. The establishment and subsequent management of national oil company Den norske stats oljeselskap (later Statoil and today Equinor ASA) was a key milestone in the history of Norwegian state participation in petroleum activities. Today, Equinor is part-privatized and listed on the Oslo and New York Stock Exchanges. The Norwegian State remains a majority shareholder.[53] Equinor no longer enjoys any preferential rights; it competes on the same terms as other oil companies for the grant of licences. However, at the time of writing it remains the dominant player on the NCS. State participation is now also enabled through direct state ownership of participating interests in licences on the NCS.

The main policies that underpinned the evolution of the Norwegian approach to state participation are much the same as those described in section 3.2.1 In

---

[53]   The Norwegian State currently holds 67 per cent of the shares in Equinor; www .equinor.com/no/investors.html, last accessed 2 April 2019.

the initial exploration phase during the 1960s, there was no direct government participation by the Norwegian State. The main rationale was the emphasis on a knowledge and experience-based approach to resource management. The need to balance the interests of the State as a host country with those of the oil companies was also acknowledged. Consequently, there was a need to offer conditions sufficient to attract foreign investors with the requisite expertise. At the time of the first licensing round, the domestic petroleum industry was nascent and the geology unknown. Further, the pioneer oil companies encountered technical risk factors due to water depths and harsh weather conditions. The conditions of the licences granted in the first round must be evaluated based on these parameters, among others.[54] When Norwegian state participation was introduced shortly thereafter, at least one of these parameters had changed: there had been discoveries of petroleum.[55] As the investment opportunities on the NCS were perceived as more attractive, there was increased scope to require state participation.

This dynamic approach to Norwegian state participation over time is also closely linked to the policy of ensuring that as large a share of the value creation as possible accrued to the Norwegian State and those aimed at building a domestic petroleum industry.[56] In a white paper entitled *St meld* nr 76 (1970–71),[57] the Norwegian government signalled that although Norway did not yet possess sufficient domestic industry expertise and there was thus an ongoing need to rely on foreign resources, the aim was to ensure that Norwegians would gain sufficient competence in all disciplines of the petroleum industry.[58]

This section presents an overview of milestones in the evolution of state participation at the different stages of development of the petroleum industry, from the very beginning to the present day.

---

[54] White Paper *St meld* nr 76 (1970–71), '*Undersøkelse etter og utvinning av undersjøiske naturforekomster på den norske kontinentalsokkel m.m*', p19.

[55] Ibid. Although the 1963 Act did not address state participation directly, it established a discretionary basis for the King to grant licences for exploration for and exploitation of petroleum. The Ekofisk field was discovered in 1969 and production started in 1971. With the 1972 Decree, express regulation of state participation was introduced. Section 31 provided that the Ministry may require state participation as a condition for grant of a production licence.

[56] See section 3.2.1. Local content development was also a part of this; see section 3.5.

[57] Ibid.

[58] Ibid.

3.4.2   State Participation Agreements (and Standard JOA) as a Condition for Grant of Licence

When state participation was introduced on the NCS, there were no express statutory provisions regarding state participation.[59] State participation was therefore first introduced as a condition for the grant of production licences. To obtain a production licence, the licensee had to enter into what was then referred to as a 'State Participation Agreement' (SPA). Initially, the conditions set for state participation in these SPAs took various forms. However, in the early 1980s a standardized SPA emerged.

The SPA was the predecessor to what today is known as the Standard JOA.[60] The Norwegian authorities still require that licensees enter into the Standard JOA as a condition for the grant of production licences. In essence, the Standard JOA is the legal instrument that establishes the rights and obligations of licensees that are jointly granted a production licence, so as to enable them to jointly carry out petroleum activities. In an international context, it is unusual that a host country requires licensees to enter into such an agreement as a condition for grant of a production licence. In many, if not most jurisdictions, the JOA is negotiated between the licensees (or, as the case may be, contractors), with little or no interference from the host country.[61] However, this practice must be seen in connection with the continued Norwegian emphasis on establishing and maintaining tight host government control of resource management and petroleum activities. This policy is also reflected in the power of the Norwegian authorities to require that applicants cooperate in joint ventures assembled by the authorities – that is, to enter into 'forced marriages' – when granting production licences.[62]

The practice of making entry into an SPA (and later the Standard JOA) a condition for grant of licences raised a debate regarding the legal status of SPAs. This debate focused in particular on whether SPAs should be regarded

---

[59]   See n 67.

[60]   The SPA was standardized in the early 1980s and is today known as the Standard JOA.

[61]   For a more in-depth discussion, see Gormley, Ovcharova and Pereira, 'To What Extent Should a Host Government Interfere in the Drafting and Conclusion of a Joint Operating Agreement?' *UEF Energy Law Review* (2016) 1.

[62]   In the invitation to tender, the MPE reserves the right to require applicants to participate in joint ventures in order to grant the licence. Oil companies have different strategies, specialities and budgets. By composing groups of licensees with different strengths and competencies, the host country can thus exercise control of petroleum activities not only through orders and regulations, but also through commercial and physical participation. Cf Arvid Frihagen, '*Statsdeltakelsesavtalene – utvikling og Standardisering*', p118.

as a condition for grant of a licence under public law or rather as private law agreements. The classification has implications for the interpretation and construction of SPAs.[63] Today there is a consensus that the Standard JOA is a condition for grant of a licence, but is also an agreement governing the *inter partes* relationship between the licensees.[64] Consequently, whether the Standard JOA will be interpreted and construed based on public law or private law considerations depends on whether the issue at hand is between the licensees and the authorities or between the licensees themselves.

By entering into an SPA, licensees accepted that this agreement regulated both the commercial cooperation and operations between the licensees, and the relationship between the State and the licensees. In effect, by accepting grant of the licence, the licensees also accepted that the national oil company was granted certain preferential rights.[65] Once grant was accepted, the licensees were not (and still are not)[66] allowed to amend the agreement without the MPE's express consent – even though the MPE was and still is not a formal party to the agreement.[67]

3.4.3 Initial Phase (1962–71)

3.4.3.1 The first licensing round
As explained in section 3.2.3, the initial phase of petroleum activities on the NCS was regulated by the 1963 Act and the 1965 Decree. The 1963 Act and 1965 Decree did not expressly regulate state participation. However, the petroleum authorities had legal grounds to require state participation through their discretionary power to grant production licences. No state participation was required in the first licensing round. In total, of 78 blocks granted, only

---

[63] Knut Kaasen, '*Statsdeltakelse i norsk petroleumsvirksomhet: Kontraktrettslig form, konsesjonsrettslig innhold eller omvendt*', *TFR* 1984 pp372–411 on the debate over the legal status of SPAs. In the early phases there were some indications that the Standard JOA had to be regarded as an agreement rather than a condition for grant of a licence, but it was subsequently agreed that the Standard JOA is rather one of the conditions of the licence.

[64] Ibid.

[65] See sections 4.4.3 to 4.4.6.

[66] The MPE's express consent is required to make changes to the Standard JOA. Cf Article 6 of the standard exploration licence; www.regjeringen.no/en/find-document/dep/OED/Laws-and-rules-2/Rules/konsesjonsverk/id748087 (text in Norwegian), last accessed 12 April 2019.

[67] The Norwegian regulatory framework for petroleum activities is intrinsically intertwined with general constitutional and administrative law. Thus, even though the Norwegian authorities enjoy broad discretionary powers, constitutional and administrative law serves to protect against discrimination, expropriation, arbitrary decisions and abuse of authority.

9 per cent were granted to Norwegian interests; the remaining 91 per cent were granted to foreign companies. Adequate government take was secured, among other things, through payment of royalties on potential production.[68] Under Section 26 of the 1965 Decree, licensees were obliged to pay a 10 per cent royalty on the gross value of petroleum produced on the site of production. This was also apparent from the invitation to tender published in *Norsk Lysningsblad* (the *Norwegian Gazette*) prior to the first round.[69]

3.4.3.2 The first State Participation Agreements

In subsequent licensing rounds, which took place from 1969–71, the Norwegian authorities introduced state participation. At this point in time – due to the results of the first licensing round, among other things – investment on the NCS had become more attractive. The Norwegian authorities were therefore able to set conditions that were more favourable to the State compared to those of the first licencing round.[70]

In the second licensing round, approximately 15 per cent of the blocks were granted to Norwegian interests (entities) and the first SPAs were negotiated. As a result of this approach, the conditions for state participation varied in these early agreements.[71] By 1969 two different types of SPAs had been entered into with four different companies and negotiations were ongoing with a fifth.[72]

In the first SPA for Production Licence 023–26 granted to the Petronord Group,[73] the State had an option to participate in development. This first agreement only addressed state participation as such and did not address operational matters. Under this agreement, the Norwegian State was given the option to become a part-owner through a one-off 'carry forward' arrangement. The State's carried interest was not limited to the exploration phase, but also

---

[68] Preparatory works NOU 1983: 16 *Organiseringen av statens deltakelse i petroleumsvirksomheten*, p6 and *St Prp* nr 36 *Eierskap i Statoil og fremtidig forvaltning av SDØE*, p19 (Norwegian preparatory works and white paper on ownership and the management of the State Direct Financial Interest).

[69] Arvid Frihagen, '*Beregning av royalty* (The point of reference for royalty value)', *studier i oljerett Institutt for offentlig retts skriftserie* 1979: 3, p12.

[70] See section 3.4.1.

[71] Moreover, in the licensing rounds between 1965–71, certain Norwegian companies were granted a number of different privileges. Two Norwegian oil companies, Hydro and Sage, were awarded privileges in relation to the foreign oil companies. Cf Arvid Frihagen, '*Statsdeltakelsens utvikling og standardisering*' *forelesninger i oljerett* 1982 p121.

[72] *St meld* nr 76, p19.

[73] The group included a group of French companies and was led by the partly state-owned Elf and Norwegian chemical company Hydro.

covered development and production.[74] A number of other models for state participation were subsequently agreed – that is, there were agreements under which:

- the State was entitled to a certain percentage of the net value in case of a commercial discovery;
- the State had the option to become a participant in the licence upon declaration of commerciality;[75]
- the State required a certain percentage of the profit margin under the net profit agreements; and
- the State had the option for the state oil company to increase its licence share upon declaration of commerciality.[76]

As the Petronord SPA did not address operational matters, the Norwegian petroleum authorities started work on drafting a more comprehensive SPA in 1969, to ensure that they – and later the national oil company – had a high degree of influence within the joint venture group and the possibility to participate in joint venture decision making, while adhering to the established international practice of using JOAs as a legal instrument to regulate the relationship between licensees in a joint venture.[77] The SPA was thus developed to include detailed provisions on the rights and obligations of operators and non-operators through the management committee,[78] in addition to special provisions on state participation.

Although state participation has to some degree been formally regulated through the SPAs, they were drafted by the Ministry based on negotiations and consultation with major foreign oil companies. There was never any public debate or parliamentary discussion regarding the particulars of the first SPA.[79]

In contrast to the approach taken in this initial phase, today the regulation of petroleum activities is relatively standardized through the 1996 Act, the regulations issued thereunder, the standard production licences and the Standard

---

[74] Arvid Frihagen, '*Statsdeltakelsens utvikling og standardisering*', *forelesninger i oljerett* 1982 p127.
[75] NOU 1983: 16 p6 and Arvid Frihagen, '*Statdeltakelsesavtalene- utvikling og standardisering*', p134.
[76] Ibid.
[77] Arvid Frihagen, '*Statdeltakelsesavtalene- utvikling og standardisering*', pp27 and 130.
[78] The SPA for Production Licence 037 was modelled on a traditional US-oriented JOA and set the pattern for later agreements; cf ibid, p132.
[79] Ibid, p27.

JOA, all of which are intrinsically interlinked.[80] There is very limited negotiation in the grant of licences on the NCS. The potential implications of allowing for more extensive negotiations in this regard are illustrated by the Supreme Court judgment in the *Phillips* case.[81] In 1965 Phillips Petroleum and two other companies were granted Production Licence 018 – one of the first licences on the NCS. The licence was granted on the basis of the 1963 Act and the 1965 Decree. One of the conditions of grant was payment of royalties which, according to the 1965 Decree, should be paid in semi-annual instalments. However, the timeframe for payments was subsequently amended under the 1972 Decree. The Phillips group filed suit against the Norwegian State, claiming damages for loss of interest. Based on the interpretation of the wording and the negotiations prior to grant of Production Licence 018 and enactment of the 1965 Decree, the Supreme Court concluded that the amended timeframe for payment could not be imposed on these specific licensees. The Supreme Court implied that the Norwegian State had entered into an agreement – and not merely granted a public law licence – with the licensees as part of the first licensing round, and that the process of the first licensing round and the 1965 Decree imposed certain restrictions on the State against subsequently changing certain fiscal parameters, such as the royalty provisions and the timeframe for payment. This represents a deviation from the main rule under Norwegian constitutional and administrative law whereby the authorities – subject to certain conditions – may unilaterally amend regulations and decrees at their discretion.

Another example of the approach taken in the initial phase is Production Licence 023–26, entered into between the Norwegian State and the Petronord Group.[82] The agreement was negotiated between the Norwegian Oil Council and A/S Petronord, which was the applicant for the licence on behalf of the Petronord Group. The agreement was called a 'Protocol', in order to indicate its preliminary form. Prior to grant of the licence, this Protocol was amended by the MPE without prior consultation with the members of the Petronord Group. This indicates that the MPE's intention was not to enter into a traditional agreement. However, there is no doubt that the Protocol was legally binding, as the key provisions were so explicit that they had to involve some degree of legal obligations which had to be subject to interpretation. In White Paper No 95 (1969–79), the Petronord Protocol was presented to Parliament

---

[80]   In a white paper submitted to Parliament in January 1973 on future participation in oil and gas developments, the first point highlighted by the Ministry of Industry was that the SPAs should be standardized to facilitate public control; cf Arvid Frihagen, '*Statdeltakelsesavtalene- utvikling og standardisering*', p130.

[81]   Published in Rt 1985, p1355.

[82]   Ibid, p127.

as an 'agreement' that should be replaced with a more detailed agreement at a later point in time. It is interesting to note that the Petronord Protocol was signed prior to grant of the licence, but was still directly linked to the licence, which had nearly been finalized at that time. One may question what the effects of this Protocol might have been had the authorities decided not to grant the licence. Under Norwegian case law, there is a need for a clear basis to conclude that the State has entered into an agreement with a private party in a context where administrative powers would normally be used. If the State in such cases chooses to enter into an agreement instead of another form of legally binding disposition, the basis for this should be evident.[83] If the State had not granted the licence to the Petronord group, there might have been a good legal basis for the Petronord Group to argue that the State was nonetheless bound by the Protocol or, more likely, liable towards the Petronord Group.

In conclusion, the grant of licences and the conditions for entering into SPAs during the initial phase were not as clear-cut as they are in present-day licensing rounds, as they involved a significant element of negotiations. It is further apparent that less host government control was exercised in this initial phase than in later phases. The *Phillips* case and the Petronord example also demonstrate that the introduction of negotiations in the award process may – even under a licensing regime – affect a host country's power to unilaterally amend certain framework conditions for petroleum activities. These examples also show that negotiations – while often necessary to attract investors to nascent petroleum provinces – can lead to (possibly) unintended results. Host countries that incorporate negotiation into the process for the award of exploration and production rights would therefore be well advised to limit negotiations to pre-defined elements only and conduct them based on a pre-defined process.

3.4.4 Second Phase (1972–78): Establishment of Statoil

One of the most important instruments under the Norwegian regime for state participation is direct and indirect ownership.[84] In the early years of Norwegian state participation, the foundations for a national oil company were built. As mentioned in section 3.2.1, one of the Ten Oil Commandments was to establish a national oil company to manage the State's commercial interests and ensure an expedient relationship with national and international partners.[85]

---

[83] RT 1929.
[84] White Paper *St Prp* nr 36 p20, item 3.6.
[85] White paper *St meld* nr 76 (1970–71).

The Norwegian national oil company – Den norske stats oljeselskap,[86] later known as Statoil and today called Equinor – was established by unanimous parliamentary decision on 14 June 1972. Statoil's main objective as national oil company was to engage in petroleum exploration and 'ensure direct influence on operations and at the same time be able to generate extensive "know-how"'.[87] Moreover, during the parliamentary debate, it was decided that Statoil would be a tool for promoting Norwegian industrial policy in the petroleum sector.[88] Another important objective of Statoil was to secure revenues from the petroleum sector for the State.[89] The company's articles of association provide that its objectives are to participate in, and cooperate with other companies towards, the exploration, production, transportation, refining and marketing of petroleum and petroleum-derived products, as well as other business.[90] This includes both traditional upstream and downstream activities, and has laid the foundation for Statoil to evolve into a fully integrated oil company.

Under Section 31 of the 1972 Decree, the Ministry had the power to require state participation[91] as a condition for the grant of production licences. This provision was linked to the newly established Statoil, though it could include other national oil companies if multiple national oil companies were established.

By unanimous parliamentary decision on 26 May 1973, the Ministry was granted power of attorney to transfer the SPAs to Statoil. At the same parliamentary session,[92] Parliament decided – by 87 votes to 23 – that all future SPAs should be entered into between Statoil and the other licensees according to further provisions to be laid down by the Ministry.[93] Through this arrangement, Statoil became a party to the existing SPAs.

With the licensing round in 1973 and the grant of the Statfjord licence, state participation entered a new phase. The SPA used in the 1973 licensing round was the first standardized SPA and Statoil was granted a licence share of 50 per cent in every block in the licensing round. Moreover, Statoil was not

---

[86]   In English, 'The Norwegian State's Oil Company'.
[87]   White paper *St Prp* nr 113 (1971–72), p8.
[88]   Preparatory works NOU 1983:16, *Organisering av statens deltakelse i petroleumvirksomheten*', p14.
[89]   Ola Mestad, *Statoil og statleg stryring og kontroll*, Marius No 105, Oslo 1983 p3.
[90]   Innst S 381 (1973–74). Statoil's articles of association were passed and subsequently approved by Parliament in 1974.
[91]   As part of state participation, Statoil was awarded certain preferential rights through provisions in the early SPA regarding a carried interest in the exploration phase, options to increase its share upon a declaration of commerciality and so on.
[92]   *Innst S* 279 and *St for* (1972–73) s 3326.
[93]   Arvid Frihagen, '*Statsdeltakelsesavtalene – utvikling og standardisering*', p27.

required to pay expenses incurred in the exploration phase (carried interest); these expenses were paid by the other participants. This mechanism was introduced instead of the arrangements in the net profit agreements discussed above. It was also included in some subsequent SPAs made applicable to other Norwegian companies. Thereafter, the carried interest was limited to the State Direct Financial Interest (SDFI).[94]

Although the SPA made Statoil a licensee, the company was granted certain other privileges in addition to being carried. Statoil was also given a special role in the procurement of goods and services with threshold values of between NOK 500 000 and NOK 10 million for licences where Statoil was not the field operator. This gave Statoil a dominant position in relation to the procurement of goods and services, as the rules entitled Statoil to participate in the preparation of negotiations and the assessment of offers. If the operator and Statoil reached different conclusions on an offer, Statoil's assessment was decisive.[95] Most decisions under the SPA were determined by simple majority.[96] As Statoil was granted a 50 per cent share in all licences, this gave it a potentially dominant position in most licences. In this phase, Statoil was also able to increase its share to 51 per cent upon expiration of the exploration period if it issued or accepted a declaration of commerciality.[97] These mechanisms allowed the Norwegian authorities to control petroleum activities not only from the outside, through the function of resource management, but also from inside the joint venture. Thus, international oil companies operating in Norway had less freedom to carry out petroleum activities as they saw fit than they were accustomed to enjoying in other jurisdictions.[98]

In 1974, when the ninth licensing round took place, a sliding scale was introduced, whereby the Norwegian State was involved in the approval of the plan for development and operation, and the State, through Statoil, had the option

---

[94] White paper *St Prp* nr 6, 'Ownership of Statoil and future management of SDFI', p20.

[95] Preparatory works NOU 1983:16, p13.

[96] Statoil's majority position in the licences was not without controversy and in the sixth round some decisions within the group were made subject to qualified majority in order to be valid. These mainly concerned: (1) approval of the yearly programme and budget; (2) unit developments; (3) main decisions on development and transportation, including the award of contracts in this with a value of more than NOK 100; and (4) termination of operatorship. Cf Ola Mestad, *Statoil og statleg stryring og kontroll*, Marius No 105, Oslo 1983, p14.

[97] Ola Mestad, *Statoil og statleg stryring og kontroll*, Marius No 105, Oslo 1983, p13.

[98] Knut Kaasen, TFR-1984-327 p377, '*Statsdeltakelsesavtalen i norsk petroleums-virksomhet: Kontraktrettslig form, konsesjonsrettslig innhold – eller omvendt?*'

to increase the licence share[99] up to specified levels described in the individual licence terms upon a declaration of commerciality.[100] This sliding scale was negotiated individually for each field. Based on the SPA and the sliding scale, Statoil's share of the licence was in some cases increased to 80 per cent.[101] The sliding scale arrangement was phased out in 1992 during the 13th concession round.

In conclusion, during this second phase from the late 1970s to the first half of the 1980s, the Norwegian authorities sought to build a strong national oil company by, among other things, granting preferential and specific rights. During this period, the Norwegian authorities significantly increased host country control of petroleum activities compared to the first phase.

3.4.5   Introduction of the Petroleum Act of 1985 and Reform of Statoil

In the early 1980s, preferential rights made Statoil one of the dominant companies on the NCS.[102] There was thus growing concern that Statoil had become excessively dominant. In the early 1980s several major discoveries were made on the NCS and Statoil's 50 per cent share generated substantial revenues for the company. This prompted a political debate, with some arguing that too much of the State's oil and gas revenues were run through Statoil.[103]

In connection with the enactment of the 1985 Act, it was decided to reform Statoil and establish an arrangement for managing direct ownership of the Norwegian State through production licences.[104] With the establishment of the SDFI, the Norwegian State gained direct participating interests in licences and petroleum facilities, rather than merely participating indirectly through Statoil. Under the sliding scale regime, Statoil's role included the collection of income on behalf of the State[105] – an arrangement that may be seen as a supplement to the royalty regime and as such arguably a more regulatory than commercial

---

[99]   As part of this regime, the percentage of licence share held by Statoil was increased. In case of major discoveries, Statoil's share could increase to a maximum of 66 to 80 per cent, depending on the individual terms of the licence. Cf Arvid Frihagen, '*Statsdeltaktelsesavtalene – utvikling og standardisering*', p125.

[100]   Ibid p20.

[101]   Ola Mestad, *Statoil og statleg stryring og kontroll*, p13 and NOU 1983: 16.

[102]   Due to the licences already granted to Statoil, it would become even more dominant. Cf NOU 1983:16, p13.

[103]   www.petoro.no/about-petoro/foundation, last accessed 12 April 2019.

[104]   White papers *St meld* nr 33 (1984–85) and *Innst S* 87 (1984–85). One of the main reasons behind the reform and the establishment of the SDFI was to ensure that Statoil did not gain a dominant position; see white paper *St Prp* nr 36, *Eierskap i Statoil og fremtidig forvaltning av SDØE*, p20.

[105]   Please see section 3.5.4.

function.[106] Introducing the SDFI was therefore a first step towards creating a clearer division between the State's commercial and regulatory functions and enabling Statoil to grow as a commercial oil company.

However, the participating interests granted to the SDFI through the licence rounds that took place after 1985 were still managed by Statoil. As the SDFI at this stage did not have its own organization, day-to-day management of the participating interests – including the marketing and sale of produced oil and gas – still rested with Statoil. In order to separate the cash flows from Statoil's own participating interests from those deriving from the SDFI interest, a separate governance model was adopted in 1992.[107] Statoil's instructions regarding management of the joint participating interests of Statoil and the SDFI were to maximize the return on invested capital over time and make decisions in a business-like and cost-efficient manner, while aiming for security of supply. The main goal was to maximize the value of the entire state participation.

During this phase – from the late 1980s to the early 1990s – the share reserved for the State was formally granted to Statoil, irrespective of Statoil's own share in the specific licence. In the different joint ventures, Statoil managed both its own participating interest and the SDFI share, although the State retained full ownership of the SDFI share. The income from the SDFI did not appear in Statoil's accounts and balance. As part of this arrangement, Statoil was responsible for selling the SDFI oil and handling all in-kind oil that the State required under the different licences (eg, as payment for area fees). Further, to ensure that the correct income was allocated to the State, norm prices were introduced to ensure that the oil sold by Statoil on behalf of the SDFI was sold on an arm's-length basis.[108]

The standardized SPA was also subject to comprehensive revisions and by the ninth licensing round in 1985, most of Statoil's previous privileges had been removed. Some were transferred back to the licensee and others were dropped. The use of the term 'SPA' was gradually phased out and replaced by the term 'Standard JOA'.[109] In addition to the procedures for management of the SDFI oil, Statoil was responsible for gas contracts as part of this arrangement.

In conclusion, during this phase, Statoil maintained a strong position on the NCS in its own right, while also handling the State's commercial and partic-

---

[106] Preparatory works NOU 1983:16, *Organiseringen av statens deltakelse i petroleumsvirksomheten*, p14.

[107] White paper *St meld* nr 21 (1991–92); cf white paper *St Prp* nr 36 (2000–01), pp36–37.

[108] White paper *St Prp* nr 36 (2000–01), pp36–37.

[109] Christian Fredrik Michelet, '*Bayona-seminaret 1989 – En vurdering av samarbeidsavtalen i 12B-runden*', Marius 175, p5.

ipating interests. This mixed role – which in part was a result of the State not having the capacity and competence to act commercially – meant that Statoil was not yet a fully commercial player on the NCS.

3.4.6   State Involvement in Sale of Gas from the NCS

In the 1970s and the early 1980s, major discoveries of gas were made on the NCS. Traditionally, gas from the NCS was sold by entering into long-term field depletion contracts.[110] Under all licences granted between 1973–78, Statoil led the negotiations and allocated gas sale contracts.[111]

In addition to state participation through Statoil and the SDFI, the Norwegian authorities established the Gas Sales Committee[112] (GSC) in 1986. Under this arrangement, Statoil was tasked with negotiating all agreements for the sale of natural gas from all fields on the NCS.[113] In the case of licences where Norwegian companies Saga and Hydro held participating interests, Statoil, Saga and Hydro formed the GSC, led by Statoil.

Under this regime, gas contracts negotiated by the GSC were so-called 'field neutral', in the sense that the long-term contract was first entered into before the MPE decided which field should deliver gas under the specific contract. From 1993 onwards, the MPE based its decision on advice from the Gas Allocation Committee,[114] comprised of representatives from different companies holding participating interests in gas fields on the NCS. The main purposes of this arrangement were, *inter alia*, to ensure long-term resource management and secure coordinated development of the major gas fields discovered on the NCS in the early 1980s. The rationale behind this arrangement was that the MPE should handle issues pertaining to resource management and the GSC should handle commercial aspects.[115]

The SPAs evolved to become more standardized as the Norwegian domestic petroleum industry matured. This regime was also reflected in the Standard JOA. If the State (ie, Statoil) accepted that the gas in a specific licence should

---

[110] For the Ekofisk and Frigg fields, the negotiations were led by Phillips and Elf.

[111] Ernst Nordtveit, *Between Market and Public Interests – Organisation and Management of the Norwegian System for Sale and Transportation of Natural Gas,* 2013.

[112] In Norwegian, '*Gassforhandlingsutvalget*'.

[113] Ernst Nordtveit, *Between Market and Public Interests – Organisation and Management of the Norwegian System for Sale and Transportation of Natural Gas,* 2013 pp472–474.

[114] In Norwegian, '*Forhandlingsutvalget*'.

[115] Ernst Nordtveit, *Between Market and Public Interests – Organisation and Management of the Norwegian System for Sale and Transportation of Natural Gas,* 2013, p474.

be sold jointly, the other participants were obliged to accept this under the SPA. This arrangement is illustrative of the strong state ownership, which – combined with the broadly formulated conditions for the grant of licences – led to substantial interference with the rights of other licensees. This regime was not expressly established under the law and is an example of 'governance by consent'.[116]

The introduction of the EEA Agreement and implementation of EU competition legislation led to a dispute between the European Commission and the Norwegian authorities on the legality of this arrangement under EU competition law. In 1996 the European Commission opened an investigation regarding a potential violation of the rules on illegal cooperation and price information in Article 81[117] of the EC Treaty (now Article 101 of the Treaty on the Functioning of the European Union (TFEU)). This ultimately led the Norwegian authorities to give up the arrangement in June 2001, one week before the European Commission issued its decision. The arrangement was brought to an end by 1 January 2002.

Following this process with the European Commission, major changes were introduced to the framework for gas transportation. Public sector company Gassco AS was established as the operator of the Norwegian gas transportation system and most of the previous upstream pipeline joint ventures for gas transportation were merged into one joint venture, Gassled, effective as of 1 January 2003.[118] This regime introduced a new approach whereby shippers and buyers would no longer negotiate access to this infrastructure separately; rather, access was regulated in the Tariff Regulation,[119] and a new Chapter 9 of the Petroleum Regulation[120] was issued under the 1996 Act based on standard contracts approved by the MPE.

3.4.7 Influence of the European Union and the EEA Agreement on State Participation

As part of the Norwegian authorities' negotiation of the EEA Agreement,[121] the EU and Norway came to an agreement that Directive 94/22/EC of the European Parliament and the Council of 30 May 1994 on the conditions for granting and

---

[116] Ibid, p475.
[117] This provision had a parallel provision in Article 53 of the EEA Agreement.
[118] www.gassco.no/en/about-gassco/gassled-eng/, last accessed 12 April 2019.
[119] Regulation 1724 of 20 December 2002.
[120] Regulation 653 of 27 June 1997.
[121] Norway entered into the EEA Agreement, which became effective on 1 January 1994. This agreement brought the EEA countries – Norway, Iceland and Liechtenstein – and EU Member States into the single market.

using authorizations for the prospection, exploration and production of hydrocarbons was relevant to their obligations under the EEA Agreement. As this directive was deemed relevant, it was decided that it should be transposed into Norwegian law. To implement the directive, the 1985 Act had to be amended. This led to the inclusion of the first provision in Section 8a of the 1985 Act to clearly address state participation and thus codify administrative practice on state participation since 1973:[122] 'The King may decide that the State shall participate in petroleum activities under this act.'

The directive did not impose limitations on state participation as such; however, some adjustments had to be made to the existing regime for state participation. The introduction of the directive correspondingly called for changes to administrative practice. The Norwegian authorities were still allowed to decide on the level of direct state participation through the SDFI, but Statoil was required to compete on equal terms for a participating interest in licensing rounds.[123] Another requirement under the directive also had a decisive effect on the organization of state participation and the management of the SDFI interests. Under Article 6, no 3 of the directive, there were certain limitations on the information that Statoil, as a manager of the SDFI interests, could receive in licences where Statoil also had its own participating interest.[124] This restriction on information flow led to a new management model for the SDFI interests whereby a new unit in Statoil was established to manage the SDFI interests in licences where Statoil did not have a participating interest, but participated as the manager of the SDFI interests. The reasoning behind this model was to reduce information sharing between this unit and other Statoil units responsible for Statoil's own participating interest.[125] Further to the influence of the directive, developments in the European gas markets and other international consolidation trends,[126] this process ultimately led to the partial privatization of Statoil and the establishment of a separate management company for the SDFI interests: Petoro.

In conclusion, in this phase – partly due to Norway's public international law obligations and developments within the international industry – the roles and responsibilities of the various players involved in securing indirect and direct state participation in petroleum activities were revised.

---

[122] Preparatory works *Ot prp* nr 63 (1994–95) p5.
[123] Ibid, p2.
[124] Finn Arnesen, *Statlig styring og EØS-rettslige skranker*, Oslo 1996, p203.
[125] Ulf Hammer et al, *Petroleumsloven kommentarutgave*, Chapter 11.
[126] *St Prp* nr 36 (2000–01), p10.

3.4.8   The 1996 Act and the Partial Privatization of Statoil

Like Section 8a of the 1985 Act, Section 3–6 of the 1996 Act establishes that 'the King may decide that the State shall participate in petroleum activities under this law'.

Towards the end of the 1990s – in response to developments in the European gas market, the consolidation of oil companies, a desire to internationalize the Norwegian petroleum industry and the restructuring of the management model for the SDFI interests – the Norwegian authorities recognized the need for further reform of Statoil[127] and a new regime for state participation. In a letter sent to the Statoil board of directors on 30 April 1999, the MPE asked for their thoughts on the future of Statoil and the SDFI.[128] On 13 August 1999 the board presented a memo suggesting that the company be partially privatized. On 26 April 2001 the Norwegian Parliament decided that one-third of the shares should be privatized and on 18 June 2001 Statoil was listed on the stock exchanges of Oslo and New York. As part of this process, Statoil was able to buy up to 15 per cent of the SDFI interests from the Norwegian State.

The main rationale behind this process was to ensure that Statoil and the SDFI interests would still ensure value creation, welfare and employment for the Norwegian people. Thus, while the main policy goal since the 1960s has not changed much, the measures for adequate implementation of that policy goal have evolved due to contextual changes. Ownership was regarded as a key measure in achieving this aim.[129] The main challenges were the maturity of the NCS, ownership issues within certain geographical areas on the NCS, international competition and consolidation of oil companies.[130] As part of this process, it was decided to establish a separate public sector corporation to manage the SDFI interests on behalf of the Norwegian State, instead of Statoil as previously. As part of the new management model, Petoro was founded on 9 May 2001.[131] As part of the privatization of Statoil, participating interests in 80 licences were transferred to Petoro. According to Petoro's bylaws, the main objective of the company is to maximize financial assets from the management of the Norwegian State's oil and gas portfolio, based on commercial principles.[132]

---

[127] *St Prp* nr 36 (2000–01), p10.

[128] Ibid, p22.

[129] *St Prp* nr 36 (2000–01), p5.

[130] Ibid, pp6–8.

[131] The Royal Decree of 18 May 2001 included a provision that the commercial side of the participating interest owned by the State should be managed by Petoro.

[132] www.petoro.no/petoro-aarsrapport/2012/om/om-petoro-og-sdoe, last accessed 5 August 2018.

As part of this process, it was decided that the core elements of the previous management arrangement with Statoil would be transferred to the new regime, as the main principles were known to the other companies on the NCS, buyers of petroleum from the NCS and suppliers.[133] These included elements such as: (1) management based on commercial principles through a commercial company; (2) presentation of all matters regarded as political or fundamental or with socio-economic effect at the general meeting; (3) the transfer of cash flows from the SDFI interests to the exchequer, rather than to the corporation managing those interests; and (4) division between the State's regulatory functions and its commercial interests.[134]

This led to the introduction of a new and detailed Chapter 11 in the 1996 Act on management of the SDFI interests. This arrangement had to comply with constitutional restraints pertaining to the management of the Norwegian State's affairs and state property.[135] Under Chapter 11, the main purpose of the manager of the SDFI interests is to maximize the value of those interests through their commercial management. Petoro, as the manager of the interests, is not involved in regulatory issues on behalf of the State, but manages the different participating interests in the same manner as other oil companies on the NCS. The main principles from the previous model were retained; so, for example, the Norwegian State holds all shares of the company, which is managed on behalf of the State – meaning for the State's account and risk. The manager, Petoro, is registered as the formal licensee for the SDFI and the payment is directly transferred to the exchequer. Petoro has the authority to vote in the different licences on behalf of the State. Statoil is still responsible for selling petroleum belonging to the SDFI interests in the international markets.[136]

---

[133] *St Prp* nr 36 (2000–01), p80; cf preparatory works *Ot prp* nr 48 (2000–01) Chapter 2.2.

[134] Ibid.

[135] Ibid, item 8.3.1 Under Section 3 of the Constitution, the government is responsible for the State's commercial affairs. Under Section 19 of the Constitution, the government may instruct and control the management of the State's property. There are also certain limitations on borrowing money on behalf of the State.

[136] Hammer et al, *Kommentarutgave til Petroleumsloven,* Chapter 11, item 1.2. Under Section 10 of Statoil's articles of association, Statoil is responsible for selling and marketing oil from the SDFI interests under instructions adopted by the company's general assembly. All oil and natural gas liquids produced under the SDFI interests are sold to Statoil and all natural gas is allocated together with Statoil's gas in a joint portfolio at the Norwegian State's risk and expense. Petoro, as the manager of the SDFI interests, shall ensure that the petroleum is allocated to the market with the best return. See www.petoro.no/petoro-aarsrapport/2017/styrets-%C3%A5rsberetning/%C3%A5rsberetning-2017, last accessed 5 August 2018.

In contrast to other licensees, Petoro does not participate in the licensing rounds. Licences granted to the SDFI (Petoro) are made on the basis of Sections 3–6 and 11–1 of the 1996 Act, but participating interests are awarded directly by the State without application.

Further, Section 12 of the Petroleum Regulations includes rules on Petoro's procedures for voting in the licence. According to Section 12, Petoro or any other company acting as manager of the SDFI interests shall use its vote 'based on visible, objective and non-discriminatory criteria'. The voting procedure shall not hinder decisions from being based on commercial assessments.[137]

Petoro is financed through allowances from government budgets and all capital income from the different licences is transferred to the State (cf Section 11–2 of the 1996 Act). Moreover, Petoro does not take on the role of operator in any licence; nor is it competent to transfer, buy or swap participating interests.[138] Under Section 11–4, Petoro is not allowed to borrow without the consent of the Norwegian Parliament. As the manager of the SDFI interests, Petoro shall present all important matters to the general meeting (ie, the Minister of Petroleum and Energy).

Under the current regime, the Norwegian State, through the SDFI, has a limited number of privileges. The SDFI is granted licences irrespective of the submission of applications and is granted new licences based on the MPE's discretion as part of the tender round.[139] Under the current version of the standardized Norwegian licence documents, the State through the SDFI is awarded pre-emption rights upon transfer of the licence interest.[140]

Through this last phase, Statoil has become a fully commercial player both on the NCS and internationally. Meanwhile, the State has maintained its direct participation through an entity that has different capacities from those of Statoil.

3.4.9 Summary of Observations

For those seeking to learn from the Norwegian experience, a key observation is that the Norwegian authorities and Parliament have actively and continuously evaluated the state participation regime, including Statoil's role and relative

---

[137] Section 12(a) of the Petroleum Regulations. This provision also includes an exemption for information on procurement of services. Petoro receives information and votes on the procurement of services only to a limited degree.

[138] Ibid.

[139] Petoro (SDFI) is directly awarded a licence in the different licence groups as part of the grant of new licences.

[140] JOA, Article 23; the pre-emption right is based on market price – see section 3.4.5.

powers. Reforms have been carried out where needed to adapt to changing circumstances. This approach has been crucial to the success of the state participation regime.

The state participation regime has evolved from no direct participation in the first licensing round to participation through Statoil and subsequently participation through the SDFI, managed by a separate corporation. As an inexperienced oil and gas nation, Norway emphasized the gradual development of a strong national oil company – Statoil – to secure revenues for the State and develop Norwegian petroleum competence and know-how. However, the Norwegian authorities were conscious that there was a need to balance the interests of the State against those of potential investors. It was only after initial exploration led to discoveries that the Norwegian authorities saw an opportunity to establish more favourable conditions for the State, including conditions on state participation.

Participation through Statoil was initially another vehicle for the State to supervise and control the petroleum industry. Through state participation, the Norwegian authorities could influence the development of the petroleum industry not only as a regulator, but also as a licensee, through voting rights in joint ventures. As the petroleum industry evolved and Statoil grew stronger, care had to be taken to clarify roles and responsibilities and maintain the balance of powers between the different players on the NCS. Hence, under the current regime, direct ownership under the SDFI interests yields the most profitable contributions to the total Norwegian government take from petroleum activities.

Stability and predictability are vital to attract foreign investment. Foreign companies can be particularly sensitive to host government interference with rights granted to them. It appears that the Norwegian authorities have managed to strike a balance between stability and reform in order to ensure that Statoil did not gain excessive dominance. Moreover, care was taken to establish terms for state participation before licences were granted. Therefore, even though the SPAs were ultimately drafted by the authorities, this practice proved acceptable to the industry.

One may ask whether, in the early days, increased regulation of state participation might have increased perceived levels of stability and predictability for investors. We have observed that the Norwegian petroleum regime and administrative law give the Norwegian authorities extensive discretion. However, Norwegian constitutional and administrative law provides legal safeguards against abuse of power, which may have reduced the need for detailed provisions on state participation. However, this must be seen in the Norwegian context, which has been characterized by a low level of political risk and stable petroleum policies over time.

On the issue of legislative development, it is interesting to see the impact of international obligations on national legislation. In the Norwegian context, the impact of the EEA and the sudden exposure to the EU and its competition laws illustrates how a host country must sometimes balance its various policy goals. Each host country must also carefully evaluate its own context when considering reforms; the 2001 reform of the state participation regime was tailored to the Norwegian context and may not be suitable in other jurisdictions. The reasons for the reform were complex and were also influenced by ongoing processes with the EU and adaption to new developments in the relevant markets. This regional free trade influence was closely linked to regional competition law concerns. The ability to adapt this regime to new developments, such as competing to attract investments in the global arena, while at the same time preserving well-functioning principles from the previous model of Statoil's management of the SDFI which were well known to the industry on the NCS, seemed to be a good solution.

Finally, while the Norwegian state participation regime has succeeded partly due to the Norwegian authorities' sensitivity to developments on the NCS, other parameters – such as the fiscal regime – have also been of great importance.

3.5 NORWEGIAN LOCAL CONTENT[141]

3.5.1 Introduction

Over the decades, Norway has successfully developed solid Norwegian petroleum competence and capabilities which have created added value for Norwegian society ('local content development').

In the early 1960s, Norway had no competence or capabilities within the petroleum industry. Today, the petroleum sector is Norway's largest, measured in terms of value added, government revenues, investments and export values.[142] The petroleum industry employs many Norwegians, directly and indirectly. There is a plethora of Norwegian oil service companies, many of which operate internationally. Several Norwegian oil companies are operating on the NCS. As demonstrated in section 3.4, the national oil company has

---

[141] There is no undisputed or universal definition of the term 'local content' in petroleum activities; however, the term typically refers to the added value that petroleum activities may bring to a host country other than the direct revenues obtained through, for instance, the sale of government shares of petroleum produced or the taxes, service fees or other dues collected from oil companies carrying out upstream petroleum activities.

[142] See www.norskpetroleum.no/en/, last accessed 3 August 2016.

enjoyed tremendous growth, developing from a wholly state-owned company that benefited from preferential rights to participate in petroleum activities to a listed, part state-owned, integrated company which is highly specialized in offshore petroleum activities. It is not only the dominant player on the NCS, but also active in over 30 countries at the time of writing.[143]

Developing local content is a goal for most host countries, but experience shows that it is not easy to achieve. This section examines the Norwegian context, local content policies and legislative measures taken to develop local content. This is largely a historical exercise. As a direct consequence of Norwegian accession to the EEA Agreement, the provisions aimed at promoting local content in the then applicable petroleum law[144] were repealed once the EEA Agreement entered into force. Thus, today there are no express local content requirements in the Norwegian petroleum regime. Norwegian players in the industry compete on the Norwegian marketplace on the same terms and conditions as international players.

This section aims to identify the reasons for the success of local content development in Norway. These are not obvious at first glance. As observed by Heum:

> In fact, the way Norway has organized the petroleum activities, the political ambitions that have been pursued, and the measures taken to implement policy are in no way significantly different from what other countries with rich endowment of oil and gas have done and attempted. Nevertheless, Norway is still one of the few exceptions when it comes to really being successful in this respect.

3.5.2 Baseline and Context

Experience shows that local content policies are more likely to achieve their intended goals if they build on realistic expectations and on existing industrial competencies and capabilities in the host country. In this respect, success might be easier to come by for a host country that already possesses some industrial competencies and a reasonable level of education in the population. Such was the case for Norway.

As discussed in Chapter 2, in the early 1960s Norway lacked the skills and competencies specific to the petroleum industry. However, Norwegian industry – including offshore and shipping – was already well developed. The design and implementation of local content policies are complex administrative and practical tasks. The Norwegian public administration was already

---

[143] www.statoil.com/no/where-we-are.html, last accessed 25 April 2017.
[144] Section 54 of Act 11 on Petroleum of 22 March 1985 ('the 1985 Act'). The 1985 Act has since been replaced with the 1996 Act.

characterized as having able, stable democratic institutions that were accustomed to promoting sound business practices, encouraging industrial dynamics and shaping industrial policies,[145] with a relatively high level of predictability, transparency and accountability. This was an excellent base on which to build a domestic petroleum industry. Internationally, the offshore petroleum industry was also in its infancy. There was therefore considerable scope for entrepreneurs to develop new technologies and establish themselves not only on the Norwegian market, but also internationally. It is hard to describe this timing as anything other than very fortuitous.

3.5.3 Key Policy Elements

Norwegian local content policies are based on the general petroleum policies discussed in section 3.2.1. The Ten Oil Commandments include express requirements to develop a new industry and to establish a national oil company. Since the early 1960s, the 'Norwegianization' of the Norwegian petroleum industry through, *inter alia*, preferential treatment of Norwegian goods and services and state participation has been an important policy goal in Norwegian petroleum resource management. To get the industry started and ensure its efficient operation, it was necessary to cooperate with experienced international companies. Heum notes that:

> politicians were in principle open to the contribution from foreign firms, while being quite determined to use the opportunity to develop domestic industrial competence. This was partly done by requiring the oil companies to set up fully operating subsidiaries in Norway, where the Norwegian authorities encouraged the recruitment of Norwegians.[146]

It was acknowledged that even from a short-term perspective, the policy goal of 'Norwegianization' had to be balanced against other policy goals – in particular, the desire to attract competent international companies and capture economic benefits from their activities. A 1979 white paper[147] sums up the main policy position on the involvement of international companies:

> It has been a clearly stated goal that the petroleum industry to the greatest extent possible shall be integrated in Norwegian business life and create economic growth and employment in Norway. At the same time it has been clear that the industry has

---

[145] Per Heum, 'Local Content Development – experiences from oil and gas activities in Norway', Institute for Research in Economics and Business Administration, Bergen, February 2008.

[146] Ibid 9.

[147] NOU 1979:43.

been completely dependent on foreign capital and expertise, and that it also in the future will be necessary and appropriate with cooperation with foreign industry.[148]

There are limitations on how far a host country can push local content policies without impairing the investment climate. Excessively rigid local content requirements or requirements that are hard to comply with may harm the interests of investors. They can also impair efficiency and increase the costs of petroleum activities, and thus ultimately also hinder the possibility of maximizing state revenues from the petroleum industry. The Norwegian authorities adopted a dynamic approach to the implementation of local content policies, conscious of the broader effects they might have on Norwegian society. For instance, in preparing the 1985 Act,[149] the drafters were contemplating the adequacy of the then applicable local content requirements as embedded in the 1972 Royal Decree. The aforementioned white paper from 1979[150] states:

> When modelling statutory provisions in this area, we face the question of whether the current rules on use of Norwegian goods and services shall be maintained unchanged or be changed... One must also keep an eye on the effects that a preferential arrangement on the Norwegian continental shelf will have for the incidental cost development in the country. A preferential arrangement will affect the prices that the industry sets for its goods and services. In turn, this can lead to an increase in the salary levels in these businesses and in the industry in general. Businesses that do not enjoy preferential treatment but that are exposed for the full effects of international competition could struggle to cope with this higher salary level. It is hard to fully foresee the consequences for Norwegian economy of such a development.[151]

This quote illustrates the dynamic approach to local content development that the Norwegian authorities adopted. It also illustrates the difficult and complex conflicts of interest and priorities that policymakers must balance, and the need to consider possible long-term effects. Institutional cooperation among relevant ministries and other state entities is therefore important if the host country is to succeed in designing and implementing adequate local content policies. When a host government has set the promotion of local content objectives as a policy goal, the petroleum sector cannot be viewed in isolation. Rather, local content policies must be in line with the other economic development policies of the host country and vice versa.[152] Sharing of information and coordination among ministries are key elements for successful implementation. Such

---

[148] Office translation of relevant parts of NOU 1979:43, p38.
[149] Act of 22 March 1985 Number 11, now repealed.
[150] Office translation of NOU 1979:43, p44.
[151] Office translation.
[152] Tordo, Warner, Manzano and Anouti, 'Local content policies in the oil and gas sector', a World Bank Study, 2013, pxiii.

coordination can be problematic in practice; communication and coordination between ministries can suffer from power struggles and turf wars. This may hinder the successful development of local content in the sense of adding value to society. In Norway, this has been much less of a problem than in many other countries, due to its well-established administrative practices[153] and the policy goal of establishing strong, competent and able institutions, among other things.

3.5.4 Legal Requirements

The goal of building industry knowledge was achieved through the implementation of different measures. As mentioned, there are no local content requirements as such in the present-day legal framework for petroleum activities. Things have thus come full circle: in the early 1960s there were likewise no local content requirements in statutory law. Local content development was instead promoted and encouraged first through 'gentlemen's agreements' between the Norwegian authorities and licensees, and later through licence conditions and specific agreements between the Norwegian authorities and licensees.

In 1972, it was decided that there should be three Norwegian oil companies operating on the NCS.[154] These companies should operate independently of each other and in the early years all three enjoyed preferential rights in order to establish themselves in the industry.

In the 1970s licences and certain additional agreements established a range of requirements to build industry knowledge. For instance, licensees were required to train Norwegian personnel to facilitate employment, government officials to help them build able institutions and educators on petroleum-related topics in Norwegian schools to build competence in relevant educational institutions. Technology transfer requirements were introduced and licensees were required to carry out a certain percentage of their research and development (R&D) in Norway. For example, the 1979 licences included a requirement that at least 50 per cent R&D activities undertaken in connection with any petroleum activities under a licence be performed in Norway. The details were set out in separate agreements. The aim was to promote R&D and develop the competence and know-how of Norwegian research institutions and contractors. This established the foundations on which to build strong research com-

---

[153] The frequent use of public hearings on new legislation has ensured that information is shared and opinions gathered before adoption of new legislation.

[154] Statoil (which was fully state owned in the beginning), Hydro (which was part state owned and part privately owned) and Saga (fully privately owned).

munities in Norway, which in turn encouraged innovation in the Norwegian petroleum industry and helped to develop local content. The Norwegian petroleum authorities today note on this topic:

> The development of new knowledge and technology is a cornerstone of Norway's resource management. . . New technology has been essential for achieving the greatest possible value creation and environmentally sound utilisation of resources on the Norwegian continental shelf. . . The industry's competitiveness and innovation capacity have also led to major positive spin-off effects and technological applications in other industries in Norway. Technology developed for the Norwegian shelf has given the Norwegian supplier industry a competitive advantage on the global stage.[155]

Local content requirements on the use of Norwegian goods and services were introduced into Norwegian law by the Royal Decree of 1972.[156] Section 54 required licensees to make use of Norwegian goods and services to the extent that these were competitive in terms of quality, service, delivery times and price. Whether this was deemed to be the case was based on a thorough overall evaluation in each specific context. Moreover, in order to enable Norwegian suppliers of goods and services to compete in tenders, Section 54 also required the inclusion of Norwegian suppliers in invitations to tender, insofar as the goods and services required were available in Norway. In evaluating tenders, the degree of local content was to be taken into consideration. Under Section 54, licensees were responsible for contractors' and sub-contractors' compliance with these requirements. Importantly, Section 54 also required licensees to carry out their petroleum activities through a base in Norway with the capacity to perform all activities and take all necessary decisions in connection with such activities.[157]

The rules established in Section 54 were carried forward in the 1985 Act. Section 54 of the 1985 Act stated that 'In activities that are comprised by this act, competitive Norwegian suppliers shall be given real opportunities to deliver goods and services'.[158] This meant that tenders and tender procedures should not make it 'unnecessarily difficult' for Norwegian entities to compete for contracts.[159] Further requirements with regard to the procurement of goods

---

[155] www.norskpetroleum.no/en/environment-and-technology/petroleum-related -research-and-development, last accessed 4 September 2016.

[156] Section 54 of the Royal Decree of 8 December 1972.

[157] In addition, there was a requirement to bring petroleum ashore to Norway. This requirement was introduced to ensure that processing and refinement would be carried out in Norway. However, this was practical only for oil, as gas has never been a significant element in the Norwegian energy mix.

[158] Unofficial office translation.

[159] *Ot prp* nr 72 (1982–83), p87.

and services were adopted in regulations. As well as promoting the involvement of several Norwegian oil companies on the NCS, these regulations helped to gradually increase the involvement of Norwegian suppliers of goods and services in petroleum activities.

The 1985 Act established further requirements on the training of personnel and provided a legal basis for the King to adopt rules on the training of public officials. This latter requirement was expanded through regulations to encompass the training of officials from the MPE, the NPD and other public entities pursuant to agreement. Moreover, it was established that the MPE could, within the ambit of the 1985 Act and Norwegian administrative law, direct licensees to train educators teaching petroleum-related topics in Norwegian schools.

Norway did not introduce specific supervision or enforcement mechanisms for compliance with local content requirements. The general rules on supervision and enforcement of the petroleum regime applied, although a specific office was established within the MPE for local content development. However, the petroleum regime and the manner in which it was implemented encouraged both prospective and existing licensees to come up with proposals as to how to ensure compliance. For instance, how prospective licensees planned to enhance local content would be considered as part of the process for the grant of new licences. Moreover, if commercial discoveries were made, a local content plan was required as a part of the development plan. In this manner, licensees were forced to be proactive in complying with local content requirements.

As noted, local content development in Norway took place rather swiftly and the Norwegian authorities were sensitive to the effects of continued preferential treatment. As Norway moved towards EEA accession in the early 1990s, a white paper[160] on, among other things, repeal of the local content requirements from the Petroleum Act highlighted the possible consequences of repeal:

> Within the petroleum sector, Norwegian oil companies have had certain reliefs and benefits compared to international companies. Thus, only the international companies have been obliged to cover exploration costs for state participation in petroleum activities. When the state has increased its shares in commercial discoveries, international companies have reduced their share, something which Norwegian companies in some cases have not been obliged to do. It is pointed out. . . that these forms of differential treatment now has ceased ... It is pointed out that ... the purchase of goods and services will be regulated in the EEA rules for public procurement. Consequentially, purchases will as a main rule take place on the basis of competitive tenders and equal treatment. . . It is pointed out that this will expand

---

[160] Innst O (1992–93).

the market opportunities for Norwegian goods and services to the petroleum industry in other EEA countries.[161]

Whether Norway would have maintained the local content requirements in
the legislation had Norway not acceded to the EEA Agreement is difficult to
comment on.

3.5.5   Summary of Observations

Norway has benefited from the local content policies it has implemented.
Results materialized very swiftly. The baseline elements outlined in section
3.5.2 were fundamental in enabling the Norwegian authorities to design and
implement adequate local content policies for the Norwegian context. They
also enabled rapid local content development. A 1979 white paper made the
following comment on local content after 14 years of petroleum activities on
the NCS:

> Over the 14 years that have passed, the Norwegian involvement has grown to be
> relatively significant. Norwegian yards have developed and built advanced drilling
> platforms and Norwegian rig owners have today a large fleet of semi-submersible
> drilling platforms, in total approximately 30 units. Norwegian yards have built
> a number of steel and concrete platforms for production and processing and large
> parts of Norwegian industry has been involved as suppliers of goods and services.
> The Norwegian participation in production licenses has been gradually increased.
> This has fortified the connections with the Norwegian business life at large.[162]

The keys to the successful development of Norwegian local content thus
perhaps lie more in the framework conditions, the timing and the manner in
which policies and requirements evolved in concert with the development of
the Norwegian petroleum industry, rather than in the policy goals and the local
content requirements themselves.

The discussion above illustrates how local content development policies
have been carefully balanced against other interests, and how policy and
requirements have been adapted in a dynamic manner. This is perhaps one of
the greatest challenges for new petroleum countries, as it requires coordination
and cooperation between various state entities. It also requires a long-term
perspective. Successful local content development requires efforts from all
main stakeholders in the industry: government, international companies and
local entrepreneurs. Host governments that aim to develop local content must

---

[161] Unofficial office translation.
[162] NOU 1979:43, p38.

therefore look beyond the ambit of petroleum law and regulation; success may require measures that address broader governance issues.

3.6 NORWEGIAN GOVERNMENT TAKE

3.6.1 Introduction

The Norwegian government take is the State's net cash flow from the petroleum industry. It includes revenue from taxation (petroleum tax, company tax, carbon tax, royalties and area fees) and other income from the SDFI and dividends from shares in Equinor. In line with the policy of maximizing value creation, government take has always been a main focus of the Norwegian authorities.

In the early days, the authorities could not have predicted the influence that the petroleum industry would end up having on the Norwegian economy. The revenues from the Norwegian petroleum sector in 2019 are expected to have increased compared with those from 2018. The total revenues from the petroleum industry for 2019 are estimated at NOK 285.8 billion (including an estimated NOK 156.1 billion in tax, the SDFI interests estimated at NOK 105.9 billion and dividends from Equinor estimated at NOK 16.6 billion, plus various fees).[163]

This section presents a brief overview of the main elements of government take, other than those derived from indirect and direct state participation.

3.6.2 Norwegian Petroleum Taxation

In the initial phase, licensees were required to pay royalties on production. Section 26 of the 1965 Decree established an obligation for licensees to pay a 10 per cent royalty of the gross value of petroleum produced at the production site. This provision has been amended and developed over the years. Under Section 4–10 of the 1996 Act, all licences with plans for development and production approved by the MPE after 1 January 1986 were exempted from the payment of royalties. Hence, some of the older fields approved prior to 1 January 1986 still pay royalties; whereas all fields with an approved plan for development after this date are not subject to payment of royalties.

Norway has successfully maintained a stable and predictable tax regime over time. In 1965 Norway introduced special regulations for the taxation of upstream activities. The objective of the new regime was to ensure that petro-

---

[163] www.norskpetroleum.no/en/economy/governments-revenues/#dividends, last accessed 22 April 2019.

leum activities were subject to the general Norwegian taxation regime, in order to increase government take.[164]

The first petroleum tax legislation was introduced in 1969. The Petroleum Tax Act of 11 June 1969 was based on the same principles as the general taxation regime. The main purpose of the act was to establish that income from petroleum activities on the NCS was subject to tax in Norway.

By the early 1970s, oil prices were high and the prospects for oil recovery on the NCS were good. This was an opportunity for Norway to secure a higher government take. At this point, government take consisted of the tax regime and royalties,[165] which were deductible under the petroleum tax regime and the company tax regime, as well as 'carry-forward' arrangements for Statoil during the exploration phase and a sliding scale in relation to state participation. In addition to these elements, as part of the mechanisms in some of the earlier SPAs, government take was also secured through the 'net profit interest' clauses, which ensured the State a share of the profits of the field subject to the SPA based on the cash flow from the specific licence.[166] It was against this background that Norway introduced its first specialized petroleum taxation regime in 1975, through the Petroleum Taxation Act 35 of 13 June 1975.

The general idea was that the oil companies shall have a 'reasonable profit' from their activities, while ensuring that the State receives a fair share of the revenue. In drafting the Petroleum Taxation Act, the following principle was highlighted:[167]

> When an assessment is made regarding which share of the production result should accrue to the government, the natural starting point is that petroleum produced on the continental shelf is exploitation of natural resources which is the property of the state.[168]

However, the Norwegian authorities also recognized that the long-term future of the Norwegian petroleum industry depended on the continued presence of multinational companies. By 1974 the Norwegian authorities had signalled their desire to establish a long-term policy regarding state participation, while recognizing the need for long-term cooperation with international companies, thus ensuring that the conditions for such activities remained attractive to those companies. This approach is still applied today.[169]

---

[164] Samuelsen, *Petroleumsskattesystemet – utvikling og veien videre*, p435.
[165] These varied between 8 and 16 per cent in the different licences; cf Samuelsen, p435.
[166] Ibid.
[167] Preparatory works *Ot prp* nr 26 (1974–75); cf ibid, p435.
[168] Ibid.
[169] Ibid.

The new regime introduced a 25 per cent special tax in addition to the ordinary company tax, which was 50 per cent at the time. Under this regime, favourable depreciation regulations were introduced, allowing the oil companies to depreciate exploration costs over six years upon production, in addition to other favourable depreciation rules which helped to make investment in the Norwegian petroleum sector attractive to oil companies. The regime was not in line with the principles of neutrality, as the value of the oil companies' investment increased on an after-tax basis, which meant that developments that were not initially profitable before tax became profitable on an after-tax basis. This regime was therefore not fully coherent with financial principles.[170]

In the early 1980s rising oil prices led to an increase in the petroleum tax rate from 25 per cent to 35 per cent, which constituted a considerable increase in the tax burden for oil companies. During this period, the average tax rate touched 85 per cent. Despite this increase, the Norwegian authorities still assessed the attractiveness of the NCS as satisfactory.[171]

From the mid-1980s onwards, falling oil prices led to new amendments to the petroleum taxation regime. As major developments such as the Troll field risked being shut down, the rate of petroleum tax was reduced to 30 per cent and a number of favourable depreciation rules were introduced, with the effect that less of the oil companies' income was subject to taxation, and new field developments were exempted from royalties. The previous 'carry forward' arrangements in the exploration phase were also scrapped. At the time, these amendments were regarded as the most comprehensive reform ever of the petroleum tax regime.

In 1987 a new provision[172] on consent from the Ministry of Finance for transactions involving participating interests was introduced, to ensure that each such transfer was effected on a tax-neutral basis.[173]

As part of the general tax reform in 1992, the company tax rate was reduced to 28 per cent and the petroleum tax rate was increased from 35 to 50 per cent. This reflected a new trend of lower company tax rates throughout Europe. Some of the amendments had a negative impact on the development of the Troll field, which had made its decision to operate based on the arrangements in place in 1986. In 2002 a number of other amendments were made to the

---

[170] Samuelsen, p435.

[171] Some of the previous very favourable depreciation rules were also amended; ibid, pp436–437.

[172] Section 10 of the Petroleum Taxation Act.

[173] Ibid, p437.

taxation regime – for example, it became possible to carry forward previous deficits without limitation.[174]

In 2003 and 2004 falling oil prices and a more negative valuation of the resource base on the NCS led to decreased exploration activities, which in turn meant less chance of new discoveries and more production. In response, a new amendment to the Petroleum Taxation Act was introduced in 2004[175] to encourage exploration activities, reduce the difference in taxation between new entrants and existing licensees and simplify and improve the framework for transactions.[176] The main measure was the introduction of a tax credit for exploration costs. This amendment offered the possibility for smaller companies that were not in a position to pay tax to benefit from deductions for exploration costs, which previously had to be carried forward. This was part of a policy that sought to ensure that newer companies on the NCS could compete with established companies on more equal terms. Further, a number of other incentives were introduced[177] which have meant that exploration activities on the NCS[178] are now less risky than those elsewhere, as companies can choose either to have 78 per cent of deductible exploration costs disbursed on a yearly basis or to carry forward the deductions when they are in a position to pay tax. Hence, companies on the NCS will only risk 22 per cent of the costs in the exploration phase. If a company decides to close down business on the NCS which is subject to the special petroleum tax regime, it may require disbursement from the Norwegian tax authorities of the tax value of any uncovered deficit for the year in which the business is closed down.[179]

Even though these adjustments have been successful, the new regime is still debated from time to time. For instance, in August 2017 non-governmental organization Bellona filed a complaint with the European Free Trade Association (EFTA) Surveillance Agency (ESA) against the Norwegian authorities. Bellona argued that the tax credit constituted state aid under Article 61 of the EEA Agreement[180] and the ESA Guidelines on State Aid for Environmental Protection and Energy. It claimed that this tax scheme had discriminatory effects towards investments in renewable energy. However, on

---

[174] As there is no ring fencing on the NCS, a deficit in field development may be used as a deduction against other income falling within the scope of the Petroleum Taxation Act.

[175] Effective from 2005.

[176] Samuelsen, p439.

[177] The Norwegian petroleum tax regime has introduced mechanisms such as favourable linear depreciation rules for facilities and a deduction for uplift as part of the tax basis.

[178] Samuelsen, p440.

[179] Section 3(C) of the Norwegian Petroleum Taxation Act.

[180] Based on an identical provision in Article 107 TFEU.

20 March 2019, the ESA concluded that the annual cash refund of the tax value of petroleum exploration costs did not constitute state aid in its preliminary examination of this case on 20 March 2019.[181]

The current tax rate for exploration and production companies operating on the NCS is 78 per cent, consisting of 23 per cent corporate tax and 55 per cent special tax.[182]

3.6.3 Other Fees

Other contributions to government take include the area fee and environmental taxes. The fees in this category have different objectives from the tax regime: they are designed to influence companies' conduct rather than to maximize the State's revenues from petroleum activities.

The area fee has been part of the Norwegian regime since the initial phase and until 1974 it was also mentioned in the licence documents. Since the 1972 Royal Decree,[183] it has been an established requirement in the petroleum legislation.[184] Section 4–10 of the 1996 Act mirrors the previous corresponding provision in Section 26 of the 1985 Act. The area fee is calculated based on the area of the licence and is linked to the duration of the licence.[185] The objective behind this fee is to ensure that the companies on the NCS are incentivized to explore and develop discoveries in an efficient manner. If licensees find that the area under licence is not commercial, this fee incentivizes them either to transfer it or hand it back to the State.[186] The current estimate for 2019 is that the area fee will constitute around NOK 7.2 billion in total.[187]

Environmental requirements are an integral part of the regulatory framework on the NCS.[188] This is also reflected in the taxes which constitute part of the Norwegian government take. Environmental taxes such as the carbon tax[189] and the nitrogen oxide (NO_x) tax are levied on players in the petroleum

---

[181] www.eftasurv.int/press--publications/press-releases/state-aid/esa-finds-no-state-aid-in-petroleum-exploration-refunds, last accessed 22 April 2019.

[182] These tax rates apply for the tax year 2018. In recent years the tax authorities have kept the overall tax rate at the same level, although they have increased the special tax rate by 5 per cent and reduced the company tax rate by 5 per cent.

[183] *Ot prp* nr 72 (1982–83), p53.

[184] Section 39(6) includes a special provision for licences awarded under the 1965 Royal Decree.

[185] Section 39 of the Regulation to the Petroleum Act.

[186] White Paper *St meld* nr 38 (2003–04) *om petroleumsvirksomheten*, p21.

[187] www.norskpetroleum.no/en/economy/governments-revenues/#taxes, last accessed 22 April 2019.

[188] The Norwegian petroleum industry is also part of the carbon trading scheme.

[189] Act Relating to Carbon Tax on Petroleum Activities.

sector. Norway was the first country to introduce a carbon tax in 1990.[190] For 2018, the tax rate is NOK 1.06 per standard cubic metre of gas or per litre oil or condensate. Under the NO_x tax regime, there is an exception for businesses which have entered into the NO_x Fund, a cooperation agreement for the industry.[191] It is estimated that environmental taxes collected in 2018 will total NOK 5.6 billion.

3.6.4   Summary of Observations

For those seeking inspiration from the Norwegian context for the development of an adequate government take model, the approach taken by the Norwegian State has aimed to balance long-term consistency with the flexibility to make necessary amendments when required, and to balance the reasonable return required by companies with the rightful revenues expected by the State as resource owner. As a result, licensees have accepted high tax rates since the mid-1970s.

For host countries, especially those with poor economies, it will always be tempting to increase tax rates when oil prices increase. However, this will affect oil companies' assessment of fiscal stability in those countries when making investment decisions. This is unfortunate, as companies typically carry out exploration and development at their own risk. Investments are huge and even if commercial discoveries are made, there is a lengthy lead time between first exploration and production. Thus, host countries that revise their government take model or parameters according to fluctuations in oil price are likely to experience reduced appetite to invest or demand stability clauses or other investment guarantees.

One of the main reasons why Norway has remained attractive to investors over time is its relatively stable and reasonable tax regime, despite changes in circumstances. The Norwegian authorities have had an incentive to keep this stable because of their long-term approach to the petroleum industry. Rather than implementing rushed, ad hoc measures to get the most out of producing fields, this approach has been reflected in a continued desire among oil companies to explore and develop new fields.[192] Further, there may be a recognition that 'as the NCS has become more mature with fewer new fields in line for

---

[190]   Ibid.

[191]   www.skatteetaten.no/en/business-and-organisation/duties1/environment/nox -tax/, last accessed 5 August 2018.

[192]   Osmunden, 'Time Consistency in Petroleum Taxation', *The Taxation of Petroleum and Minerals*, p455.

development, the government will depend on a reputation as a predictable and reasonable tax collector to avoid under-investment'.[193]

Norway has successfully used the tax regime to encourage investment and make necessary adaptions for continued encouragement. It has been able to maintain a consistent high-level government take policy throughout different governments and Parliaments. Norway has further transitioned from field-based solutions and adaption to a generic system based on the principle of neutrality in line with the general Norwegian tax regime. Based on economic theory, a high tax rate on the resource rent will have less impact, as the Norwegian special tax only targets extraordinary profit. It is thus less likely to have an impact on companies' assessments, with undesired effects.

Under the Norwegian petroleum tax regime, there have been few cases of indirect expropriation, as most amendments have applied only to new licences. Further, in a number of decisions relating to both petroleum tax and other sectors, the Norwegian Supreme Court has made it clear that the principle of legality stands firm.[194] Hence, the amendments described above have seemingly had limited impact on companies' perceptions of the stability of the Norwegian petroleum tax regime.

However successful this approach has been in the Norwegian context, those seeking inspiration from the Norwegian experience must take into consideration that the Norwegian petroleum tax regime is complex, advanced and built around the general Norwegian tax regime, which involves significant tax administration. The taxation system has evolved over time and may not necessarily be transferred to other jurisdictions without adaption.

3.7 REVENUE MANAGEMENT

Revenue management typically poses a challenge for host countries. As new discoveries were made in the early 1980s, the idea that a certain portion of the revenues from the petroleum sector should be placed in a fund gained traction.[195] The idea behind this fund was that only the marginal rate of return should be used as part of the government budget; the remaining fund should be a 'buffer' for future government finances.

In 1990 the Norwegian Government Pension Fund (also known as the Norwegian Sovereign Wealth Fund)[196] was established through Act 36 of 22 June 1990 (the 'Pension Fund Act'). The main mechanism for transfer to

---

[193] Osmunden, 'Time Consistency in Petroleum Taxation', *The Taxation of Petroleum and Minerals*, pp455–456.

[194] Including Supreme Court Decisions Rt-2004–1921 and Rt-201–143.

[195] Preparatory works, NOU 1983:27.

[196] The fund is also known as the Norwegian Oil Fund.

the fund relates to income based on cash flow from the petroleum industry transferred from the annual budget, yield from the assets of the fund and net financial transactions in connection with petroleum operations.[197] The cash flow includes all tax and royalty revenue under the Petroleum Tax Act and the 1996 Act; revenue from the carbon tax; income from the SDFI interests; income from the State's net profit agreements; share dividends from Equinor; transfers from the Petroleum Insurance Fund; and proceeds of sale of SDFI assets. The following costs are subtracted: the State's direct investments in the petroleum industry; the State's costs incurred in connection with the Petroleum Insurance Fund; the State's costs in relation to decommissioning and disposal of facilities; and any potential costs in connection with acquiring assets as part of the SDFI activities.[198] According to the preparatory works to the Pension Fund Act, the fund will increase when the cash flow from the petroleum sector exceeds the sum transferred from the fund to the annual budget and reduce when the cash flow is less than the sum transferred to the annual budget.[199] Based on the mechanism and fluctuations in oil price, the first transfer of funds from the petroleum sector was made by the Ministry of Finance in 1996.[200] This was the first year since the fund was established that a positive cash flow based on this mechanism and fluctuations in oil price was generated. The 1990 Act was subsequently amended by Act 123 of 21 December 2005.

The main objective of the Government Pension Fund is stated as follows: 'The Government Pension Fund shall support government saving to finance the National Insurance Scheme's expenditure on pensions and support long-term considerations in the use of petroleum revenues.'[201]

Currently, the Government Pension Fund has an estimated value in excess of NOK 8 000 billion. As of the amendment to Section 2 in 2009,[202] the fund is divided into two parts: the Government Pension Fund Global[203] and the Government Pension Fund Norway. The Ministry of Finance is in charge of managing the Government Pension Fund Norway, while the Government Pension Fund Global is managed by Norges Bank Investment Management

---

[197] Section 2 of Act 36 of 22 June 1990.

[198] Ibid.

[199] Preparatory works, *Ot prp* nr 29 (1989–90) *Lov om statens petroleumsfond*, item 1.1.1.3.5.3.

[200] *St Prp* 1 (1996–97) Annual Budget, Chapter 2.5.2.

[201] Section 1 of Act 123 of 21 December 2005.

[202] Amendment of 18 December 2009.

[203] For further information on the governance model for the Government Pension Fund Global, see www.nbim.no/en/the-fund/governance-model/, last accessed 18 February 2018.

subject to guidelines issued under the Investment Management Mandate.[204] According to the investment strategy, the Government Pension Fund Global has invested in international equities, fixed income and real estate. Further, the fund is subject to extensive supervision and internal and external audits in order to ensure compliance with applicable policy and legislation.[205]

As mentioned above, use of the fund is limited to transfers to the annual budget pursuant to a resolution of the Norwegian Parliament.[206] As part of this resolution, a practical application has emerged known as the 'spending guideline' ('*Handlingsregelen*'). This rule of thumb was established through an understanding between a majority of the political parties in the Norwegian Parliament, and holds that Government Pension Fund spending should be limited to the expected annual net real return of the fund, which at that time was expected to be four per cent.[207] This has since been reduced to three per cent. The general idea behind this arrangement is that income from the petroleum sector should be gradually phased into the Norwegian economy.[208] These budget guidelines are flexible and are adapted to the economic cycle each year.

3.7.1 Summary of Observations

While the Government Pension Fund has been a great success in the sense of its magnitude, its objectives were specially tailored to the Norwegian context at the time of its establishment. The Government Pension Fund Global has exceeded expectations due to developments in revenues, fluctuations in oil and gas prices and the management of the fund. Parliament's decision to restrict the spending of the fund has also contributed to its growth. Whether other host countries can successfully introduce a similar fund will depend on state finances and whether all revenues from the petroleum sector are needed as part of the annual budget. The first transfer to the Government Pension Fund was made six years after its establishment. The level of transparency and

---

[204] The mandate for management of both the Government Pension Fund Global and the Government Pension Fund Norway is further stipulated in separate regulations issued under the act.

[205] The scope of this chapter does not allow for extensive presentation of the supervisory mechanisms of the fund. However, an illustration of these mechanisms is available at www.nbim.no/en/the-fund/governance-model/supervision/, last accessed 18 February 2018.

[206] Section 5 of Act 123 of 21 December 2005.

[207] Ola Mestad, 'Managing the Nation's Wealth – the Government Take and the Resource Curse', SIMPLY 2014 s 211-226 MARIUS-2014-456-211.

[208] www.statsbudsjettet.no/Statsbudsjettet-2018/Statsbudsjettet-fra-A-til-A/ Handlingsregelen/, last accessed 18 February 2018.

accountability apparent in its management, operation and investment rules can nevertheless serve as inspiration for other host countries.

3.8   NORWEGIAN INVESTMENT PROTECTIONS AND OTHER INCENTIVES

3.8.1   Investment Protection

The core of investment protection in Norway lies in the generally applicable constitutional and administrative law. In case of expropriation, the Norwegian Constitution requires that full compensation be paid to investors.[209] Moreover, the administrative law sets out requirements for due process. Although the Norwegian legal regime affords wide discretionary powers to the MPE, there are limitations as to how the MPE may exercise such powers. Such discretionary powers must be exercised within the ambit of many statutory and case-based principles of due process, such as non-discrimination, protection against expropriation, protection against arbitrary decisions and protection against abuse of authority.

At the international level, Norway has ratified certain treaties that include provisions on investment protection, such as the EFTA free trade agreements with Singapore[210] and Ukraine.[211] Norway has also implemented the European Convention on Human Rights (ECHR), which protects rights to property.[212] Moreover, the EEA Agreement provides investors from the EU/EEA with protection against discrimination and the freedom of establishment.[213]

One issue that has been hotly debated in recent years is the use of bilateral investment treaties (BITs). Although Norway has not entered into any BITs since the mid-1990s, the government has started work on forming a mandate for the negotiation of new BITs, in order to support Norwegian companies' investments abroad by allowing them to compete on equal terms with other foreign investors.[214]

Norway entered into 14 bilateral investment treaties before the mid-1990s.[215] The substantive protections included in those BITs vary to a certain degree, but

---

[209]   Cf 105 of the Norwegian Constitution.

[210]   In 2002.

[211]   In 2012.

[212]   ECHR Protocol I, Article 1.

[213]   EEA Agreement, Articles 4, 31 and 40.

[214]   This policy aim was reconfirmed in the platform for the enlarged centre-right government formed in January 2018.

[215]   Norway is currently a party to BITs with Chile, China, Czech Republic, Estonia, Hungary, Latvia, Lithuania, Madagascar, Peru, Poland, Romania, Russia, Slovakia

most provide for the right to fair and equitable treatment, protection against expropriation, most-favoured nation treatment and investor-state dispute settlement. For foreign investors based in contracting states, these Norwegian BITs add an extra layer of protection. Norway has also signed the Energy Charter Treaty (1994) (ECT), but has not ratified it due to constitutional concerns.

Norway is a party to the 1958 New York Convention on Recognition and Enforcement of Foreign Arbitral Awards, as well as the 1965 Washington/ International Centre for the Settlement of Investment Disputes Convention.

The Norwegian judiciary is regarded as non-biased. A limited number of cases between the State and investors or licensees have come before the Norwegian courts over the years. One major case that has challenged this view more recently is the legal action against the Norwegian State by four foreign investors claiming that an amendment to the tariff regulations was invalid and that the State was liable for damages.[216] The Norwegian State prevailed in the Supreme Court on 28 June 2018. The Supreme Court found that the amendments to the tariff regulations were valid (see further Chapter 4).[217]

The Norwegian courts are generally considered to be genuinely independent and resistant to pressure from the other branches of government. Norway is ranked as the world's sixth least corrupt country by Transparency International.[218]

3.8.2   Summary of Observations

The Norwegian legal system affords robust protection to foreign investment. The Norwegian Constitution contains solid protections of ownership, as well as protection against retroactive legislation. The Norwegian courts are by and large effective and independent of the government, and corruption is low. The failure to ratify the ECT is a weak point, which precludes investors in the Norwegian energy sector from a course of legal action available to investors in many other European countries. The political culture is stable and trust in institutions high. In sum, investment protection in Norway is strong.

---

and Sri Lanka; www.regjeringen.no/no/tema/naringsliv/handel/nfd---innsiktsartikler/ frihandelsavtaler/investeringsavtaler/id438845/, last accessed 23 October 2018.

[216] For more details of the matter, see www.regjeringen.no/en/topics/energy/oil-and -gas/lawsuit-over-gassled-tariffs/id2406034/, last accessed 22 April 2019.

[217] www.domstol.no/globalassets/upload/hret/henviste-saker/henviste-saker -internett.pdf, last accessed 24 October 2018.

[218] Transparency International, *Corruption Perceptions Index 2016* January 2017; http://issuu.com/transparencyinternational/docs/2016_cpireport_en?e=2496456/ 43483458, last accessed 24 October 2018.

3.9   CONCLUSION

It is clear from the discussion in this chapter that the successful management of Norwegian petroleum resources is partly due to framework conditions such as timing, resource base, a stable democracy, industry experience, established administrative practices, competent institutions, unbiased courts and a low level of corruption. As such, specific elements of the Norwegian model must be evaluated within the Norwegian context.

Throughout this chapter, we have illustrated how Norwegian petroleum policy and the legal framework for petroleum activities are characterized by strong host country control. We have shown how resource management is based on policies as implemented in legislation and through the exercise of administrative authority. The licensing regime is the primary vehicle for Norwegian host country control of resource management, in combination with requirements for additional permits, approvals and consents and the power to stipulate conditions for issuing the same. This is not an unusual model for resource management in modern international petroleum law; it appears that the strength of the Norwegian model is rather due to the context and the manner in which it has been implemented and developed over the decades.

Although the Norwegian model cannot be transferred wholesale to other host countries, inspiration may be taken from some of its general policies and approach to implementation. Some of these have been recurring themes throughout this chapter and include:

- a step-by-step, knowledge-based approach to the design and implementation of policies and legislation;
- a long-term perspective in designing and implementing policies and legislation;
- an acknowledgement that the petroleum industry may have a broader impact on the host country, and that therefore the development of petroleum policies and legislation may conflict with other policy goals of the host country, which may present a need to balance conflicting policies from time to time;
- a focus on promoting predictability and stability, while still making reforms where necessary due to changing circumstances and policy goals;
- a focus on balancing the needs of the State with those of the oil companies;
- a focus on good governance by strong, competent and able intuitions; and
- continued dialogue with the petroleum industry to capture changes in framework conditions, whether in respect of the market, technological developments or otherwise.

All of the above elements can be tweaked to suit most countries in the long run, although it may take considerable time and effort to make these changes. For instance, implementation of adequate levels of transparency and accountability as a part of good governance may be challenging. Poor economies and weak institutions may hinder the implementation of long-term policies and fiscal stability. Many host countries also struggle to engender trust between the State and the people, as well as between the State and the oil companies.

The Norwegian legal framework affords the authorities a considerable degree of discretionary power. Although this has proved a useful legislative tool in Norway, it is possibly one feature of the Norwegian model that may be difficult to export. Even in the Norwegian context, which is characterized by high continuity and predictability in policy and safeguards in administrative law, one may question whether the manner in which the Norwegian authorities exercise these powers is sufficiently transparent. As the powers of the Ministry are so wide, one may also question whether the systems for accountability as established are working. Companies may be reluctant to lodge complaints against such a powerful body. For instance, we are not aware of any complaints being lodged by companies that have not been awarded licences. We note, however, that recent history may indicate an increased level of lawsuits challenging the exercise of discretionary powers, including for instance the *Gassled* case. The fact that the Norwegian State prevailed at all three instances may further illustrate the challenge for foreign investors when assessing the different risk factors prior to making investment decisions. Due to the wide discretionary powers embedded in the 1996 Act and subordinate regulation, investors will need extensive knowledge of the Norwegian regime prior to entering this market.

4.   Licensing regime: Norway

Henrik Bjørnebye and Catherine Banet

4.1   INTRODUCTION

The Norwegian petroleum resource management model is based on a licensing regime set out in the Norwegian Petroleum Act,[1] with further details provided in the Petroleum Regulations.[2] The Act governs the lifecycle of petroleum operations from cradle to grave through a step-by-step licensing approach. The main considerations and principles behind the Petroleum Act have been discussed in earlier chapters of this book. In this chapter we will provide an overview of the content and structure of the licensing system, following a stepwise approach from the opening of petroleum areas to decommissioning of installations.

Upstream oil and gas activities have different impacts on their surrounding environment in the different stages of exploration, production and transformation. In this chapter we will therefore also examine the manner in which offshore petroleum activities are regulated with respect to their interaction with the surrounding environment. We will consider this in two stages. First, we will analyse how the different economic, social and environmental interests related to the use of relevant sea areas are balanced on the opening of areas for petroleum exploration. Second, we will consider the regulation of emissions from offshore petroleum activities during the production phase.

The Norwegian Petroleum Act applies only to petroleum activities relating to subsea petroleum deposits subject to Norwegian jurisdiction.[3] Since all discovered Norwegian petroleum resources are located on the Norwegian Continental Shelf (NCS), this means that the Act in practice governs all proven deposits.

---

[1]   Act 29 November 1996 No 72 relating to petroleum activities, as amended.
[2]   Regulations 653 to the Act relating to petroleum activities of 27 June 1997, as amended.
[3]   Section 1–4 of the Act.

The Petroleum Act is structured in 12 chapters, beginning with introductory provisions on fundamental issues such as the right to resources and resource management, scope and definitions (Chapter 1), and ending with entry into force (Chapter 12). The award of an exploration and production licence pursuant to Section 3–3 of the Petroleum Act is a key decision in this system. This licence provides the licensee with an exclusive right to exploration and production within the licence area in cooperation with the other licensees on the basis of a joint operating agreement (JOA). However, the rights under an exploration and production licence must be seen in relation to the other licence requirements in the Act, such as approval of a plan for development and operation (PDO) of the field at issue.

Section 1–1 of the Act sets out the fundamental point of departure that ownership of Norwegian subsea petroleum resources is vested in the State. Consequently, unproduced offshore petroleum deposits as such are not subject to private property rights. This means that licensees with an exploration and production licence pursuant to the Act do not acquire ownership of the petroleum resources that they exploit until the petroleum flows past the wellhead during the production process. No entity other than the State as such may carry out petroleum activities without the necessary licences or permits governed by the Act.[4] Consequently, the activities of the partly state-owned company Equinor ASA (formerly Statoil) (67 per cent owned by the State), and its subsidiary Equinor Energy AS, as well as the wholly state-owned company Petoro AS, are subject to licensing.[5]

Section 1–1 also sets out the more obvious point that the State has the exclusive right to resource management. This principle would have followed from Norwegian legislation even in the absence of such explicit provision. The Act delegates to the King in Council – that is, the government – responsibility for resource management in accordance with the provisions of the Act and decisions adopted by the Norwegian Parliament.[6] The Ministry of Petroleum and Energy (MPE) acts as resource management authority in practice, with important aspects of management delegated to the Norwegian Petroleum Directorate (NPD).

The purpose of the Norwegian petroleum legislation is described indirectly in the resource management directions set out in Section 1–2 of the Petroleum

---

[4] Section 1–3 of the Act.

[5] Petoro's management of the State's direct financial interest in the petroleum sector, as further governed by Chapter 11 of the Petroleum Act, is described in more detail below in section 4.4.

[6] Section 1–2, first paragraph of the Act.

Act.[7] This provision is relevant for the interpretation of the other provisions of the Act and is worded as follows:

> Resource management of petroleum resources shall be carried out in a long-term perspective for the benefit of the Norwegian society as a whole. In this regard the resource management shall provide revenues to the country and shall contribute to ensuring welfare, employment and an improved environment, as well as to the strengthening of Norwegian trade and industry and industrial development, and at the same time take due regard to regional and local policy considerations and other activities.[8]

In the following, we will provide an overview of how the Norwegian petroleum resource management model and its interaction with environmental considerations are enshrined in the main requirements of the Petroleum Act and related legislation through the petroleum lifecycle.

4.2 OPENING OF NEW AREAS FOR PETROLEUM ACTIVITIES

Exploration permits for seismic surveying as well as exploration and production licences can be awarded only in areas that have been opened for petroleum activities. The decision to open new areas for petroleum activities thus has major policy implications.

Not all areas that contain or are suspected to contain petroleum resources are necessarily opened for exploration and exploitation. It has been decided that certain areas of the NCS should remain closed for petroleum activities. This is mainly a political decision and the government's stance may vary according to the governing party's position on the matter. For example, the current government led by Erna Solberg is a coalition government between four political parties ('Solberg II government'). In a joint government declaration signed by the four governing parties, it was agreed not to open for petroleum activities or carry out environmental impact assessments (EIAs) in the sea areas outside Lofoten, Vesterålen and Senja in the period 2017–21, and not to commence petroleum activities near Jan Mayen, the Ice Edge, Skagerak or the Møre fields.[9] This is in line with former agreements in the *Storting* that there will

---

7   Section 1–2, second paragraph of the Act.

8   Unofficial English translation by the NPD, available at www.npd.no/en/ Regulations/Acts/Petroleum-activities-act/, last accessed 2 April 2019.

9   Political platform for the government consisting of Høyre, Fremskrittspartiet, Venstre and Kristelig Folkeparti, agreed at Granavolden 17 January 2019, p92; available at www.regjeringen.no/contentassets/7b0b7f0fcf0f4d93bb6705838248749b/ plattform.pdf, last accessed 2 April 2019. See also the similar statement in the former

be no petroleum activity off Jan Mayen, the Ice Edge, Skagerak or the Møre fields.[10] A more recent development subject to intense debate between the government, the opposition parties and scientists has been the redefinition of the Ice Edge, which has been proposed to be moved further north, with the consequence of opening up new areas that were previously excluded.

During the last 15–20 years,[11] the Norwegian government has strengthened its efforts to elaborate a more consistent strategy for the management of the marine areas under its jurisdiction.[12] This move corresponds to a general trend in terms of environmental management and is in line with Norway's international and European Economic Area (EEA) commitments. More specifically, it has resulted in the adoption of so-called 'Management Plans' for the following three marine areas: the North Sea and Skagerrak;[13] the Norwegian Sea;[14] and the Barents Sea, including Lofoten.[15]

The purpose of the Norwegian Management Plans for marine areas is to facilitate value creation, the coexistence between industries and the sustainable harvesting of resources.[16] The plans contribute to the implementation of integrated ecosystem-based management of the marine environment in Norwegian waters. Ecosystem-based management is a well-known environmental management approach which has found echoes in public international law and which has been applied to the marine environment since the 2000s.[17] It entails that the management of human activities must take as a starting point the limits set by the ecosystem itself, in order to maintain its essential structure, functioning, production and biodiversity. It looks at the whole range of inter-

---

political platform for the government consisting of Høyre, Fremskrittspartiet and Venstre, agreed at Jeløya on 14 January 2018, p71; available at www.regjeringen .no/contentassets/e4c3cfd7e4d4458fa8d3d2bb1e43bcbb/plattform.pdf, last accessed 2 April 2019.

[10] See White Papers 26 (1993–94), 37 (2008–09), and 10 (2010–11).

[11] See in particular, White Paper 12 (2001–02) *Rent og rikt hav*; *Innst S* 161 (2002–03) White Paper 19 (2004–05) *Marin næringsutvikling. Den blå åker* (*Innst S* 192 (2004–05).

[12] For an analysis of the Management Plans for Norwegian Sea areas, see HC Bugge, '*Har vi de rettslige redskapene som trengs for en god forvaltning av våre havområder?*', in M Stub and I Hjort Kraby (eds), *Forsker og formidler. Festskrift til Erik Magnus Boe på 70-årsdagen 17. april 2013* (Universitetsforlaget, 2013), pp65–87.

[13] Adopted in 2013 (White Paper 37 (2012–13)); next review in 2030.

[14] Adopted in 2009 (White Paper 37 (2008–09)); currently under review, next update in 2025.

[15] Adopted 2006; updated in 2011 (White Paper 10 (2010–11)); next update in 2020.

[16] See White Paper 37 (2008–09).

[17] The Convention on Biological Diversity is one of the central pieces of public international law in the matter.

actions within an ecosystem. In the Norwegian context, it shall facilitate the coexistence of different industries such as fisheries, shipping and petroleum operations within the relevant marine environment. The Management Plans cover waters from the baseline to the open sea, as well as human activities in those areas.

The elaboration of the Management Plans starts with an ecosystem-based assessment for each main economic activity in the area, as well as an assessment of the interactions between the relevant commercial activities, such as petroleum, fisheries and shipping. The plans also define measures to reduce the environmental burden of those activities or to solve competitive uses of the same sea area.

In the case of petroleum and energy activities, this preliminary assessment falls under the competence of the MPE, which gathers representatives from the different interest groups in a working committee in charge of drafting the assessment.[18] Basically, the zone covered by a Management Plan can encompass four main types of petroleum areas: areas in which there will be no petroleum activity; areas not yet opened for petroleum activity, but subject to an opening process (eg, Jan Mayen and the part of the previously disputed area to the west of the delimitation line in the Barents Sea South); areas where a process of opening has started; and opened areas (in both mature and frontier areas). Based on the results of the preliminary assessment, the Management Plans can, among other things: set conditions for the opening of new areas to petroleum exploration; decide to keep certain areas closed for petroleum activity or other activities; establish traffic separation systems; introduce improved safety or emergency measures; and adopt measures to secure the state of the environment. Areas not opened for exploration are subject to monitoring by Norwegian authorities. For example, the NPD collates geological and geophysical data.

Section 3–1 of the Petroleum Act requires that an impact assessment be carried out prior to the opening of new areas for petroleum activities. The MPE is the lead ministry for completion of the impact assessment. Further requirements – such as the drafting of a proposed assessment programme and hearing rounds – are set out in the Petroleum Regulations adopted pursuant to the Petroleum Act.[19] The content of the impact assessment programme and the resulting impact assessment is defined in Chapter 2a of the Petroleum

---

[18]   This assessment undertaken by the working committee under the direction of the MPE is a different process from the traditional and regulated impact assessments, such as those under the Petroleum Act.

[19]   Chapter 2a of the Petroleum Regulations.

Regulations (Sections 6b and 6c respectively). The whole process is very similar to a strategic EIA.

The opening process consequently starts with a proposal for impact assessment programme, which, if approved, is followed by completion of the impact assessment itself. The purpose of the impact assessment is to 'describe the presumed impacts of opening of the area for petroleum activities, the different possible development solutions and the impact of future petroleum activities in the area'.[20] In broad terms, when evaluating the impact assessment, the MPE will look at the environment, trade and industry, the risk of pollution, and the economic and social effects of opening for petroleum activities. In particular, the impact assessment shall include a description of:

- the area(s) planned to be opened for petroleum activities;
- the relationship to national plans relevant for the area to be opened, and to relevant environmental goals/standards laid down through national guidelines, national environmental goals, white papers and so on, and how these are reflected in the impact assessment;
- the assumed impacts on employment and commercial activities, as well as the expected economic and social effects of the petroleum activities;
- important environmental and natural resource issues;
- the impact of the opening on, among other things, living conditions for animals and plants, the sea bed, water, air, climate, landscape, emergency preparedness and risk;
- the possible transboundary effects of the opening;
- the need for, and any proposals in relation to, further investigation before opening; and
- the measures available to prevent or compensate for any possible damage or disadvantage.[21]

The opening of new areas is also subject to a hearing process which involves local public authorities, central trade and industry associations and other interest organizations.[22] The impact assessment shall also be made available to the public online. Interested parties shall be given a timeframe of no less than three months to present their views on both the impact assessment programme and the impact assessment.

If the impact assessment concludes positively, the MPE will usually propose the opening of the area for licensing. Although not explicitly set forth in the Petroleum Act, the Norwegian Parliament customarily decides whether to

---

20 Section 6c of the Petroleum Regulations.
21 Section 6c of the Petroleum Regulations.
22 Section 3–1(2) of the Petroleum Act.

open new areas for petroleum activities on the basis of the impact assessment.[23] Consequently, the government submits a White Paper prepared by the MPE to the Parliament. If it agrees, the *Storting* opens the new areas for licensing. Again, the impact assessment forms part of the White Paper, as well as the comments received during the consultation process and an evaluation thereof. In its White Paper to the *Storting*, the MPE can make further recommendations in order to take into account the specificities of the area and the interests present. The MPE shall consider in the White Paper whether the opening should be made subject to requirements for further investigations to monitor and show the factual impacts of the petroleum activities. The White Paper shall also consider whether it is necessary to set specific conditions to reduce and compensate for significant adverse effects.[24] The fact that the decision to open new areas is made by the legislative branch makes it more difficult to review legally.

Legally speaking, the decision to open new areas does not entitle any entity to exploration and production rights, and the State has full discretion to decide whether to award licences or to leave an opened area untouched. Consequently, the petroleum companies have no legal legitimate expectations that they will be awarded licences in an opened area, and a decision by the State not to award licences in such areas will clearly not raise any question of State liability. The assessment of applications under the licensing system is also subject to additional criteria relating to the capacities of the applicant and the project for development of the deposit. At the same time, the opening of areas is a strong political signal, usually rapidly followed by the inclusion of the newly opened areas in the next licensing round. Against this background, the decision to open new areas is *de facto* considered an important step in the process of commencing new petroleum activities.

The formal opening of new areas has two main legal implications. First, survey licences may be awarded pursuant to Section 2–1 of the Petroleum Act. This licence gives the holder a right to carry out geological, petrophysical, geophysical, geochemical and geotechnical activities – that is, in practice, seismic surveying. In addition, shallow exploration drillings are allowed.[25] The licence does not, however, allow for ordinary exploration drilling, which is covered by the exploration and production licence awarded pursuant to Section 3–3 of the Act. The survey licence does not give the holder an exclusive exploration right in the area covered by the licence.[26]

---

[23]   This follows also indirectly from Section 6d of the Petroleum Regulations.
[24]   Section 6d of the Petroleum Regulations.
[25]   Section 4 of the Petroleum Regulations.
[26]   Section 2–1, second paragraph of the Petroleum Act.

Second, exploration and production licences may be awarded pursuant to Section 3–3 of the Petroleum Act in areas opened for petroleum activities.[27] This is the most fundamental licence enshrined in the Petroleum Act, as it entails an exclusive right to exploration, exploration drilling and production of petroleum deposits within the area covered by the licence.[28]

4.3 THE EXPLORATION AND PRODUCTION LICENCE

The exploration and production licence is awarded by the King in Council – that is, the government – and may cover one or several petroleum blocks or parts of blocks.[29] The MPE shall announce in advance the areas for which applications for exploration and production licences can be submitted, with a minimum time limit of 90 days for filing applications.[30] A collection charge of NOK 109 000 accrues for each application filed to the MPE. Licences are granted for up to ten years, with a possibility for extension if work commitments and other conditions are fulfilled.[31] In practice, exploration and production licences are awarded through numbered licensing rounds which are typically held every second year. Following the publication of the 24th licensing round in June 2017, the MPE announced the award of new licences in June 2018. Originally, 102 blocks were proposed, based on the nominations from companies, the results of public consultation and the final evaluation by the government authorities; and 12 new licences in 47 blocks were ultimately awarded in both the Barents Sea and the Norwegian Sea. In addition to the numbered licensing rounds in more frontier areas, a practice was introduced in 1999 – later institutionalized in 2003 – of having annual awards in mature petroleum areas in addition to the ordinary licensing rounds. Those so-called

---

[27]   This licence is referred to as a 'production licence' in the NPD's English translation; but as it also covers ordinary exploration activities by way of exploration drilling, we will refer to it as an 'exploration of production licence'.

[28]   Section 3–3, third paragraph of the Petroleum Act.

[29]   Section 3–3, first paragraph of the Petroleum Act. The area opened for petroleum activities is divided into blocks in accordance with Section 3–2 of the Petroleum Act.

[30]   Section 3–5, first paragraph of the Petroleum Act. The announcement shall be made in the *Norwegian Gazette* (*Norsk Lysingsblad*) and the *Official Journal of the European Union*; see Section 3–5, second paragraph. The wording in Section 3–5, fourth paragraph may at the outset seem to open up for awards without prior announcement; but – read in conjunction with the rest of the provision, as well as the Hydrocarbon Licensing Directive – it must be understood as providing only an exemption from the ordinary 90-day application time limit; see Hammer et al, *Petroleumsloven* (Universitetsforlaget, 2009), pp146–148.

[31]   Section 3–9 of the Petroleum Act.

'Awards in Predefined Areas' rounds, modelled on the numbered licensing round system, are considered by the MPE as equally important for the development of petroleum activities on the NCS.[32]

The Petroleum Act sets forth that a licence shall be granted on the basis of 'factual and objective criteria' and the requirements and conditions stated in the notification.[33] The government has full discretion to consider whether to award licences based on the applications received.[34] The Petroleum Regulations present more detailed criteria by setting out that licences shall be awarded based on the applicants' technical competence, financial capacity and plan for exploration and production in the area comprised by the application in order to further best possible resource management.[35] The realization of the objectives enshrined in Section 1–2, second paragraph of the Petroleum Act, cited above, is also relevant for the evaluation of applications.[36] Moreover, Article 5 of the Hydrocarbon Licensing Directive, which is part of the EEA Agreement, also sets out overall criteria. The Directive emphasizes, *inter alia*, that the award criteria shall be objective and non-discriminatory, and these requirements are implemented in the Petroleum Regulations.[37] In addition, non-statutory principles of Norwegian administrative law include some fundamental legal safeguards, such as a prohibition on government abuse in making governmental decisions.

Despite the overall criteria enshrined in the Petroleum Act, the Petroleum Regulations and the Hydrocarbon Licensing Directive, it is clear that the Norwegian government enjoys a wide margin of discretion in evaluating applications and granting exploration and production licences. Unlike in some other jurisdictions, for example, awards are not based on auction. At a general level, a non-auction system with broad evaluation criteria and a wide margin of discretion for the government in taking decisions of significant economic value can raise the risk of challenges relating to lack of transparency and objectivity, and in the worst case corruption. These challenges appear not to have materialized in Norway and the Norwegian system for licence awards has generally worked according to the interests it pursues. However, whether the Norwegian system at this point would be well suited for replication elsewhere

---

[32]   *Prop* 114S (2014–15), *Norges største industriprosjekt – utbygging og drift av Johan Sverdrup-feltet med status for olje- og gassvirksomheten*, p11.

[33]   Section 3–5, third paragraph.

[34]   Section 3–5, third paragraph second sentence.

[35]   Section 10 of the Petroleum Regulations.

[36]   See similarly Roggenkamp et al (eds), *Energy Law in Europe* (Oxford, third edition, 2016) p820 at 11.50.

[37]   See Article 5 of the Directive and Section 10 of the Petroleum Regulations.

is not obvious and is something that would have to be considered carefully on a case-by-case basis.

The Petroleum Act confers competence on the government to stipulate licence requirements for the licensee as a part of the licence award.[38] These provisions supplement the general competence to stipulate conditions to a public decision that follows from Norwegian non-statutory administrative law principles. Licence requirements must be based on the need to ensure that the petroleum activities covered by the licence in question are carried out in a proper manner.[39] Moreover, the conditions can be based only on considerations for national security; public order; public health; transport safety; environmental protection; protection of biological resources and national treasures of artistic, historic or archaeological value; the safety of the facilities and employees; systematic resource management (eg, production rate or optimization of production activities); or the need to ensure fiscal revenues.[40] These restrictions on the government's discretion to set licence conditions are based on national implementation of the Hydrocarbon Licensing Directive.[41]

The licence terms have developed over time and licensing rounds, but some fundamental terms have been present throughout most of the petroleum development era, including the obligations to carry out a work programme, to pay an area fee and to enter into the Standard Joint Operating Agreement (JOA).[42] While these obligations are explicitly referred to in the Petroleum Act, the government may also impose other conditions based on its general competence to stipulate licence conditions.

It follows from Section 3–8 of the Petroleum Act that the government may impose a specific work obligation on the licensee for the licence area. This work obligation may consist of exploration and exploration drilling of wells as further specified and within the timeframe set out in the licence.[43] Fulfilment of the work obligation (as well as other licence conditions) is a condition for requesting an extension of the initial ten-year licence period.[44]

Consequently, the licence extension rules provide an incentive for licensees to carry out the work programme. Moreover, the area fees incentivize licensees to progress exploration and production activities within areas where licences

---

[38]   See Section 3–2, first paragraph as well as the general right to impose requirements as part of decisions in Section 10–18, fourth paragraph of the Petroleum Act.

[39]   Section 11, first paragraph of the Petroleum Regulations.

[40]   Section 11, second paragraph of the Petroleum Regulations.

[41]   See Articles 6(1) and (2) of Directive 94/22/EC.

[42]   Roggenkamp et al (eds), *Energy Law in Europe* (Oxford, third edition, 2016) p820 at 11.51.

[43]   Section 13 of the Petroleum Regulations.

[44]   Section 3–9 of the Petroleum Act.

have been awarded. The obligation to pay an area fee arises after the expiry of the initial licence period.[45] Exemptions are granted upon application if exploration or production activities are carried out in the area in question.[46] The fee is calculated per square kilometre and increases over the first three years of accrual.

The government sets general obligations as to the coexistence of the different economic activities in the zone and the environmental effects of petroleum exploitation. The Petroleum Act requires that petroleum activities be conducted 'in a prudent manner', and that they:

> not unnecessarily or to an unreasonable extent impede or obstruct shipping, fishing, aviation or other activities, or cause damage or threat of damage to pipelines, cables or other subsea facilities. All reasonable precautions shall be taken to prevent damage to animal life and vegetation in the sea, relics of the past on the sea bed and to prevent pollution and littering of the seabed, its subsoil, the sea, the atmosphere or onshore.[47]

The Petroleum Regulations further require, as a condition for grant of a production licence, that the petroleum activities be carried out 'in a proper manner'.[48]

In addition, the licence can set specific obligations in order to reflect the specificities of the block. Those specific obligations can relate to the effects of the envisaged petroleum activities on the environment and on other economic activities such as fisheries. Such requirements are directly attached to the block and specified in the announcement for licensing and grant of the licence.

The award of a production licence in a block does not preclude others from obtaining a licence of rights to explore the same area for the purpose of the production of natural resources other than petroleum resources or scientific research, on the condition that this does not hamper unreasonably the petroleum activities covered by the petroleum licence. If the exploration for other activities is conclusive and causes inconvenience to the petroleum activity, the King shall decide which of the activities shall continue or be postponed.[49]

---

[45] Section 4–10, first paragraph of the Act.

[46] Section 39 of the Petroleum Regulations.

[47] Section 10–1(2) of the Petroleum Act.

[48] The latter means that grant of the licence shall be conditional on: 'consideration for national security, public order, public health, transport safety, environment protection, protection of biological resources and national treasures or artistic, historic or archaeological value, the safety of the facilities and employees, systematic resources management or the need to ensure fiscal revenues (Section 11, second paragraph, Petroleum Regulations).

[49] Section 3–13 of the Petroleum Act.

4.4 THE JOINT OPERATING AGREEMENT (JOA)

The Petroleum Act expressly provides that the government may require, as a condition for grant of a licence, that licensees enter into agreements with specified contents with one another – that is, a JOA.[50] This condition is in practice always stipulated when licences are awarded to more than one entity for a defined area. Consequently, the licence groups are in practice put together by the government on the basis of individual licence applications by the companies and subsequent licence awards. The MPE appoints or approves an operator for the licence group.[51] In practice, one of the licensees is appointed operator of the group although the Act also allows for the selection of a non-licensee.[52] State-owned company Petoro, which manages the State's Direct Financial Interest (SDFI) in the petroleum sector, participates only as licensee on the NCS and not as operator of any licence groups. The partly state-owned company Equinor, on the other hand, operates as a normal licensee and is appointed operator for a large number of licence groups. As the largest commercial petroleum company on the NCS, Equinor is also the company with the largest operator portfolio.

The parties to a JOA are the licensees in each licence group. The government as such is not a party to the agreement. The standard parts of the agreement are, however, drafted by the government, which updates them regularly. These standard parts consist of the JOA standard terms as such and an accompanying Accounting Agreement. The licensees are required to enter into these terms as a licence condition and they are not subject to negotiation. The JOA also consists of a specific part for each licence; but this section only includes the name of the parties and their share of the licence, the corresponding voting rules based on shares as well as the designated operator of the licence group. All activities carried out under the licence are governed by Norwegian law, and are based on Norwegian contractual conditions and subject to Norwegian rules on dispute resolution.

Consequently, the Norwegian JOA is an agreement entered into between the entities to each licence without participation by the government as such, but with strict government control over the content of the agreement. This means that the agreement has both private law and public administrative law aspects. On the one hand, the agreement is a contract governing the rights and obligations between the licensees, and not between the licensees and the government, subject to ordinary principles of contract law. On the other hand, the fact that

---

[50] Section 3–3, fourth paragraph of the Act.
[51] Section 3–7, first paragraph of the Petroleum Act.
[52] Section 3–7, third paragraph of the Act.

the licensees are required to enter into the agreement on fixed terms as a public law licence requirement may in some cases be relevant for the interpretation of the agreement.

The overall objective of the JOA is to govern the relationship between the licensees in carrying out the petroleum activities covered by the licence – that is, the exploration and production of petroleum deposits. It consequently governs the rights and obligations between the parties through the phases of development and production, as well as development of the field, transfer of ownership interests and cessation.

By entering into the JOA, the parties establish a joint venture. This partnership shares many of the traits of an unlimited liability company. However, joint ventures established pursuant to production licences under the Petroleum Act are specifically exempted from the Norwegian Companies Act, and the joint venture is therefore not a company under Norwegian law.[53]

The cooperation under the JOA comprises petroleum production, but not petroleum sales, as it is a competition law requirement to market and sell separately. This means that the title to the petroleum produced is transferred to each licensee in accordance with its participating interest in the joint venture for individual company-based sales.[54] Internally, the licensees are liable primarily on a *pro rata* basis and secondarily on a solidary basis towards each other for commitments arising due to the activities of the joint venture.[55]

The Management Committee is the governing body of the joint venture. It decides on overall strategy, and it determines guidelines and controls the conduct of the operator. All licensees are represented on the Management Committee, with the operator acting as leader. This Committee has the power to establish sub-committees, which as a main rule have an advisory function.

The operator shall carry out the daily management of the joint venture's activities and act on behalf of the joint venture. The operator carries out these tasks on the basis of a no gain, no loss principle.[56] Thus, the operator will be liable for loss incurred by the other licensees only where it has caused the loss through gross negligence or wilful misconduct by leading personnel.[57]

---

[53]   Section 1–1 (1) of Act 83 of 21 June 1985.

[54]   The scheme for company-based sales replaced the former field-based sales scheme which was challenged by the European Commission under the competition rules of Article 81 of the EC Treaty (now Article 101 of the Treaty on the Functioning of the European Union) in the so-called *GFU* case related to the Norwegian Gas Negotiation Committee, settled by the Commission in July 2002 (COM/36/072).

[55]   Section 7.1 of the Standard JOA.

[56]   Section 3.1 of the Standard JOA.

[57]   Section 3.5 of the Standard JOA.

Norwegian petroleum policy has been based, *inter alia*, on comprehensive state participation in the licence groups since the beginning of sector development.[58] The Norwegian state oil company Statoil was established in 1972 with a view to increasing state revenue, influence and know-how.[59] From the third licensing round, Statoil was granted a 50 per cent participating interest in all licence groups.[60] Under this scheme, the JOAs included several important advantages for Statoil, such as carried interest through the exploration phase.[61] In 1985 the State's interests in the sector were reorganized into the SDFI, which included the State's direct shares in the petroleum sector, and Statoil's licence shares. Under this scheme, Statoil managed the SDFI on behalf of the State. From the 15th licence round, Statoil was no longer granted a licence share in all licences.[62]

Another important reform was carried out in 2001, when Statoil was part-privatized and thus no longer considered a vehicle for state policy. At the same time, the fully state-owned company Petoro was established in order to take over management of the SDFI. Management of SDFI is now specifically governed in Chapter 11 of the Petroleum Act, which sets out that the State reserves a share of the licence for state ownership, and that this participating interest shall be managed by a fully state-owned limited company – Petoro.[63]

Petoro is formally a licensee for the participating interests it manages on behalf of the State and acts as a party to the JOAs with the rights and duties of a participant.[64] The State is directly liable for any obligations incurred by Petoro.[65] The revenues from the management of the SDFI belong to the State.[66] However, Petoro does not market and sell SDFI petroleum. Pursuant to a marketing and sales instruction issued by the MPE, Statoil markets and sells the

---

[58] 'National supervision and control' was the first of the so-called 'Ten Oil Commandments' adopted in 1971 by the Standing Committee on Industry of the *Storting*. National control had three dimensions – central management, administrative and commercial functions – with the commercial dimension being represented at the start by a fully state-owned company, Statoil.

[59] Roggenkamp et al (eds), *Energy Law in Europe* (Oxford, third edition, 2016) p825 at 11.70.

[60] Roggenkamp et al (eds), *Energy Law in Europe* (Oxford, third edition, 2016) p825 at 11.70.

[61] Roggenkamp et al (eds), *Energy Law in Europe* (Oxford, third edition, 2016) p825 at 11.70.

[62] Roggenkamp et al (eds), *Energy Law in Europe* (Oxford, third edition, 2016) p825 at 11.71.

[63] See Sections 11–1 and 11–2 of the Petroleum Act.

[64] Section 11–2, second paragraph of the Petroleum Act.

[65] Section 11–3 of the Petroleum Act.

[66] Section 11–2, third paragraph of the Petroleum Act.

SDFI oil and gas together with its own petroleum in order to secure the highest possible value for the State's and Statoil's petroleum.

4.5   DEVELOPMENT AND OPERATION

The production of petroleum is governed by Chapter 4 of the Petroleum Act. The Chapter commences with an overall prudent production standard in Section 4–1. This standard sets out that petroleum production shall be carried out with a view to maximizing production from each deposit 'in accordance with prudent technical and sound economic principles and in such a manner that waste of petroleum or reservoir energy is avoided'.

The PDO is central to the production of petroleum. If a licence group decides to develop a deposit – that is, if exploration drilling results in a commercial finding – the licensees must submit a PDO of the deposit to the MPE for approval.[67] This plan shall include a description of all relevant aspects relating to the development – that is, economic, resource, technical, safety related, commercial, environmental, transportation and decommissioning aspects of the planned activities, including an impact assessment.[68] Consequently, the licensees must submit a detailed plan describing the production solutions, such as whether to build subsea or platform installations or use floating production storage and offloading units, whether to include processing facilities or process the wellstream at other existing facilities and so on. In order to avoid pressure for the government in deciding whether to approve the plan, the licensees cannot undertake substantial contractual obligations or start construction before the plan has been approved, unless the MPE agrees otherwise.[69] The MPE shall state the reasons for the decision to approve or not to approve a PDO in a publicly available document.[70]

The licensees must present an EIA programme to the MPE well in advance of submitting the PDO.[71] The EIA submitted as part of the PDO 'shall state the reasons for the effects that the development may have on commercial activities and environmental aspects, including measures to prevent and remedy such effects'.[72]

The licensees' proposal for an EIA programme shall contain a description of the development solutions to envisaged effects in relation to other commercial

---

67   Section 4–2, first paragraph of the Petroleum Act.
68   Section 4–2, second paragraph of the Petroleum Act and Section 20 of the Petroleum Regulations.
69   Section 4–2, fifth paragraph of the Act.
70   Section 20, fourth paragraph of the Petroleum Regulations.
71   Section 22 of the Petroleum Regulations.
72   Section 22a of the Petroleum Regulations.

activities and the environment, including possible transboundary environmental impacts. The proposed EIA programme is subject to a hearing process, after which the MPE decides on the content of the final EIA programme. The overall impact assessment to be undertaken for the development and operation of the petroleum deposit shall be prepared on the basis of the EIA. It shall therefore also cover aspects relating to environmental impact and effects on other commercial activities. The impact assessment is forwarded to the interested parties and published online for comments. The MPE takes the final decision to approve the PDO as it is or subject to amendments.

One question which may be raised concerning the relationship between the exploration and production licence on the one hand and approval of the PDO on the other is the extent to which the government retains discretion to deny the PDO when the licensees have already been granted a right to exploration and production pursuant to Section 3–3 of the Petroleum Act. While the exploration and production licence relates to the general right to develop resources, the PDO relates to the choice of development method. This means that, on the one hand, the government must clearly be entitled to refuse a PDO if the proposed development method does not satisfy relevant standards such as environmental, safety, technical or commercial standards. On the other hand, the power to deny a PDO cannot *de facto* be applied to reverse a decision to award an exploration and production licence without relevant reasons relating to the development proposal. Such decision would easily amount to a reversal of the licence decision under Section 3–3 of the Act, which would then have to comply with the general requirements for reversal of administrative decisions pursuant to Section 35 of the Norwegian Public Administration Act and non-statutory law. The threshold for accepting such reversal is high.

The PDO may also include intra-field pipeline infrastructure, such as wellstream pipelines from one installation to a processing facility on another intra-field installation. Oil extracted and processed on the field will typically be lifted and transported by oil tanker from the field to refining facilities and markets in other countries. However, in most cases natural gas will be transported in pipelines from the field to an onshore landing point. The building of such pipelines is subject to a separate licence to install and operate and a Plan for Installation and Operation (PIO) pursuant to Section 4–3 of the Petroleum Act, to the extent that this right does not already follow from the approval of the PDO. A PIO is a plan for construction, placement, operation and use of facilities for petroleum activity, including shipment facilities, pipelines, cooling facilities, facilities for production and transmission of electricity and other facilities for transport or utilization of petroleum. The construction or operation of different infrastructure may involve separate licences and PIOs. As the PDO must account for the total development concept, there may be a need for clarification of the parts of the development covered by the licences in the PDO and the parts of the infrastructure covered by the PIO. This is determined by the MPE.[73]

Infrastructure and pipelines subject to licensing under Section 4–3 of the Act are typically installations subject to third-party access, such as Gassled, which will be further examined below. Facilities approved under a PDO may, however, also be subject to third-party access – for example, a processing facility where other fields may tie in rather than invest in separate processing facilities.

Finally, the MPE shall approve the production schedule prior to approving the PDO/PIO or granting a licence pursuant to Section 4–3, and the MPE shall stipulate the quantity which may be produced for fixed periods upon application from the licensees.[74]

In some cases a petroleum deposit may turn out to extend to other blocks with other licensees or to the Continental Shelf of another State, such as the UK. Section 4–7 sets out that the licensees for the blocks involved shall seek to reach agreement on joint petroleum activities in such cases (ie, a unitization agreement). Such unitization agreements shall be submitted to the MPE for approval no later than when the PDO is submitted. The MPE may also determine how the activities are to be conducted and how the deposit will be apportioned between the licensees in the licence groups if agreement is not reached within reasonable time. A recent example was in relation to the Johan Sverdrup field in 2015, when the MPE had to reach a decision on the allocation of ownership shares between the licensees. For deposits extending to the Continental Shelf of another State, a bilateral treaty between the States involved is also necessary to determine the approach to petroleum resource management, since several States are involved. The Statfjord field in the North Sea, located on the border between the Norwegian and UK sectors, is an example of such field.[75]

---

[73]   On the relationship between the PDO and PIO and an interpretation of Sections 4–3 and 4–4 of the Petroleum Act, see *Ot prp* 46 (2002–03), pt 2.3. The MPE and the Ministry of Labour and Social Affairs have also issued guidelines for the preparation of the PDO and the PIO: Guidelines for plan for development and operation of a petroleum deposit (PDO) and plan for installation and operation of facilities for transport and utilization of petroleum (PIO), last updated March 2018.

[74]   See Section 4–4 of the Petroleum Act.

[75]   The Agreement between Norway and the United Kingdom relating to the exploitation of the Statfjord field reservoirs and the offtake of petroleum therefrom was entered into on 16 October 1979. For further details see also www.norskpetroleum.no/ fakta/felt/statfjord/, last accessed 2 April 2019.

4.6   THIRD-PARTY ACCESS TO INFRASTRUCTURE

A general point of departure is that the owner of a facility has the exclusive right to use the facility unless otherwise provided by law or agreement. In sectors with facilities that may be regarded as natural monopolies, such as the gas and electricity sectors, the regulation of third-party access on non-discriminatory, transparent and fair terms has been considered a key instrument to promote competition on equal terms and the socio-economic efficient utilization of infrastructure.

A third-party access regime can in principle be either negotiated or regulated. Under a negotiated scheme, the government typically requires that the owner of the facility open it for use by third parties on transparent, non-discriminatory and fair terms, but then leaves it to the parties to negotiate the commercial terms, which must respect these fundamental requirements. Within a regulated scheme, the commercial terms for the use of the system – that is, the transportation tariff – are regulated in laws and regulations.

The Hydrocarbon Licensing Directive focuses on exploration and production licensing requirements and therefore does not govern access to infrastructure. The relevant provisions derive from both primary EU/EEA law (competition law provisions) and secondary law provisions. Indeed, the EU gas market legislation contains extensive rules requiring a regulated third-party access scheme for gas transmission and distribution pipelines, primarily regulated in the Gas Directive (2009/73/EC) and Gas Regulation (EC) No 713/2009. The Gas Directive and Gas Regulation do not comprise field installations, however, and the Gas Directive requires only the introduction of a negotiated third-party access regime for upstream gas pipeline networks – that is, pipelines transporting gas from the field installations to onshore terminals and final landing points.[76]

Section 4–8 of the Norwegian Petroleum Act confers competence on the MPE to decide that facilities approved pursuant to Sections 4–2 (ie, PDOs) and 4–3 (ie, other infrastructure and pipelines) shall be open to third-party access. The provision distinguishes between upstream pipeline networks[77] on the one hand and other facilities on the other. For upstream pipeline networks, natural gas undertakings and customers established within the EEA area shall have a right of access to the facilities, including the necessary facilities supplying

---

[76]   The scope of the Gas Directive is set out in Article 1 of the Directive. See also the definition of 'upstream pipeline network' in Article 2(2) and the specific negotiated third-party access regime for such networks in Article 34 of the Directive.

[77]   See the definition of 'upstream pipeline network' in Section 1–6 (m) of the Petroleum Act, which is based on the corresponding definition in the Gas Directive.

technical services.[78] For other facilities, typically field facilities, the MPE may require third-party access on the basis of considerations for efficient operations or the benefit of society, and provided that such access is not unreasonably detrimental to the licensee's own requirements or those of others which already have a right of use. This approach is in accordance with EU law, which requires as a minimum a negotiated access regime for upstream pipeline networks and does not set any third-party access requirements for fields.

Third-party access to field facilities, such as processing, is governed by a separate regulation on third-party access pursuant to the Petroleum Act ('the TPA Regulation').[79] The TPA Regulation sets out the procedures for negotiating third-party access between two licensing groups and the overall principles for the agreement, including overall principles for the determination of tariffs and other commercial terms. According to these provisions, the owner is entitled to a tariff that covers operating costs, investments, compensation for lost profit such as lost and deferred production due to third-party use, other documentable losses as well as a reasonable profit taking into account the risk incurred by the owner in providing third-party access.[80] Third-party access agreements shall be submitted to the MPE for approval and any disagreement between the parties arising under the TPA Regulation may be brought before the MPE for decision.[81]

The question of third-party access to field installations may raise a number of different questions and challenges based on the merits of each case. In particular, the economic risk for the owner of a facility in granting access to a third party may vary depending on the risk that the third-party petroleum may lead to lost or deferred petroleum production for the host field. Since lost and deferred petroleum production has a high cost for licensees, considering the risks involved can be important in determining tariff levels for third parties. Consequently, it is difficult to establish regulated tariffs for third-party use of field facilities and there is much merit to a negotiated scheme such as that established under the Petroleum Act and the TPA Regulations.

For mature upstream pipeline systems, on the other hand, a regulated third-party access scheme may have more merit. The offshore gas pipeline infrastructure for the transportation of natural gas from the NCS to the UK and the European continent is the largest offshore gas pipeline system in the world.

---

[78]   Section 4–8, first paragraph.
[79]   Regulation 1625 of 20 December 2005. See Roggenkamp et al (eds), *Energy Law in Europe* (Oxford, third edition, 2016) pp836–848 for a more detailed analysis of third-party use of production facilities.
[80]   See Section 9 of the TPA Regulation.
[81]   See Section 4–8, second paragraph of the Petroleum Act and Section 13 of the TPA Regulations, correspondingly.

It is nevertheless to be considered an upstream pipeline network within the meaning of the Gas Directive and it is therefore not subject to the Directive's more comprehensive regulation of gas transmission and distribution activities.[82] Consequently, Norway has the discretion to decide whether to apply a negotiated or regulated third-party access scheme for this network system.

With very few exceptions, the offshore gas pipeline infrastructure system on the NCS, including the facilities supplying necessary ancillary services such as processing of rich gas, is owned by the Gassled joint venture. This joint venture resulted from the merger of separate pipeline joint ventures in 2003. The parties were originally the same companies as the field licensees, with ownership shares corresponding to the participating interests on the fields. In later years, other parties – including institutional investors such as pension funds – have bought shares in the Gassled joint venture, which is consequently no longer owned entirely by oil and gas companies operating on the NCS.

The upstream pipeline system owned by the Gassled joint venture is subject to a regulated third-party access regime governed in more detail by Chapter 9 of the Petroleum Regulation, as well as a separate Tariff Regulation stipulating the specific transportation tariffs.[83] The tariff levels have been decided on the basis of the long-term Norwegian policy that the value from the petroleum resources should be collected at the level of the fields, and not in the transportation system. An overall 7 per cent pre-tax rate of return on investments has traditionally been applied as a guiding principle for the stipulation of gas transportation tariffs.

The Gassled joint venture is operated by Gassco AS, which is a state-owned limited share company with no ownership interests in the infrastructure. Gassco operates Gassled on behalf of the joint venture, but at the same time is appointed to assume extended operator responsibility pursuant to Section 4–9 of the Petroleum Act. This extended operatorship entails that Gassco shall ensure third-party access on non-discriminatory and transparent terms by acting neutrally towards all gas shippers, including the owners of Gassled. In carrying out these neutrality obligations as further set out in Chapter 9 of the Petroleum Regulations, Gassco thus carries out public law obligations in addition to its private law obligations as operator for and on behalf of the Gassled joint venture. This independent system operator approach, combined with a regulated access regime, seeks to ensure grid neutrality, although some of the Gassled joint venture partners are also involved in gas production and

---

[82] Norwegian gas is exported almost in its entirety and domestic gas consumption is limited. The existence of onshore gas pipelines to be considered distribution pipelines under the Gas Directive is therefore strictly limited.

[83] Regulation 1724 of 20 December 2002.

shipping in the same system. It is beyond the scope of this chapter to provide a detailed analysis of the regulated third-party access regime for the upstream gas pipeline network.[84]

Following a consultation process, the MPE amended the Tariff Regulation in 2013 with a view to reducing tariff levels significantly for future unbooked transportation capacity. This decision was challenged before the Norwegian courts by four companies that had recently acquired participating interests in Gassled, claiming that the tariff amendment decision was invalid or alternatively that they had a right to compensation for their loss due to reduced tariff income. Both the Oslo City Court and the *Borgarting* Court of Appeal decided that the amendment was not invalid and that the claimants did not have a right to compensation.[85] Only the question of invalidity was brought before the Supreme Court, which decided that the amendment was not invalid and consequently that the government was legally entitled to amend the Regulation as it had done.[86]

4.7   CESSATION

Petroleum operations require large-scale installations that need to be handled properly after the end of operations in order to avoid negative impact on the environment and other interests such as fisheries. The regulation of cessation plans and decommissioning of installations is therefore an important part of the resource management regime, governed by Chapter 5 of the Petroleum Act and Chapter 6 of the Petroleum Regulations.

The Norwegian cessation and decommissioning regime must also comply with international law obligations which are binding on Norway; in this regard, the Convention for the Protection of the Marine Environment of the Northeast Atlantic (OSPAR Convention), to which Norway is a party, is of particular importance.[87] OSPAR Decision 98/3 on the Disposal of Disused Offshore

---

[84]   For a more detailed (although not entirely updated) analysis, see Ulf Hammer, 'System operation' and Anne-Karin Nesdam, 'Third party access to upstream pipeline networks on the Norwegian Continental Shelf' in Ulf Hammer et al, *Articles in Petroleum Law* (Marius 404, 2011), pp64–77 and pp78–128, respectively. See also Torkjel Grøndalen and Cato Lower, 'Third Party Access to Infrastructure on the Norwegian Continental Shelf', 4 *LSU J of Energy L & Resources* (2016).

[85]   Judgments of 25 September 2015 and 30 June 2017. The case is currently under appeal to the Norwegian Supreme Court.

[86]   Supreme Court, Judgment HR-2018-1258-A in Case 2017/1891 delivered on 28 June 2018, *CapeOmega AS, Solveig Gas Norway AS, Silex Gas Norway AS, Infragas Norge AS v the Norwegian State/ Ministry of Petroleum and Energy*.

[87]   The Convention for the Protection of the Marine Environment of the North-East Atlantic, signed 22 September 1992.

Installations provides that dumping disused offshore installations and leaving them in place wholly or partly at the outset is prohibited, although in certain situations the competent authority can grant permission to leave large steel installations and certain concrete installations in place.

The licensees must submit a decommissioning plan to the MPE before the end of the exploration and production licence or the licence for transportation installations pursuant to Section 4–3 of the Act. This decommissioning plan may, for example, propose continued petroleum operations, use of the facilities for other purposes, partial or full removal of the installations or abandonment. The plan must contain the information necessary for the authorities to make a decision relating to disposal.[88]

The MPE shall adopt a decision on disposal, including a time limit for carrying out the decision.[89] The licensee and owners are under the obligation to carry out the decision; and in the case of assignment of participating interests during the licence period, the assignor is still alternatively liable for financial obligations arising due to the disposal.[90]

4.8 PETROLEUM, THE ENVIRONMENT AND CLIMATE: EMISSIONS REGULATION

4.8.1 Introduction

Petroleum operations raise significant risks of negative impact on the local environment. The main sources of emissions from offshore oil and gas activities in Norway relate to produced water, use of chemicals, emissions to air, and mud cuttings and waste. The sector-specific petroleum legislation addresses these risks through requirements relating to, *inter alia*, EIAs at the various stages of petroleum operations, licence requirements, strict liability rules for pollution damage and elaborate health, safety and environment (HSE) rules based on internal control systems as the regulatory philosophy. The sector-overarching Pollution Control Act also applies to petroleum operations.[91]

The Norwegian government has adopted a policy goal of zero environmentally harmful discharges from petroleum activities, known as the 'zero discharge goal'.[92] In accordance with this goal, substances that are harmful to the environment in principle cannot be discharged into the sea. Furthermore,

---

[88] Section 5–1 of the Petroleum Act.
[89] Section 5–3 of the Petroleum Act.
[90] Section 5–3, third paragraph of the Act.
[91] Section 4 of Act 6 of 13 March 1981.
[92] This goal was adopted in 1997 and further defined in Report 28 (2010–11) to the *Storting*.

the objective is to minimize the risk of environmental harm caused by the discharge of all sorts of chemical substances into environments (air, soil). The zero discharge goal applies to all offshore operations, including drilling and well operations, production and pipeline transportation. It has served as the main policy orientation since its adoption and is reflected in the legislation.

Climate change policy sets particular parameters for operations in the oil and gas industry, which represents one of the largest sources of greenhouse gas (GHG) emissions in the world. Because Norway is a large oil and gas exporter, the reduction of GHG emissions from the petroleum sector has important consequences for the reduction of emissions of the country as a whole.

A global reduction in GHG emissions with a view to meeting the targets set in the Paris Agreement to the United Nations Framework Convention on Climate Change (UNFCCC) will require significant reductions in the production and use of fossil energy. This raises the crucial question of whether petroleum producing countries should not only endeavour to reduce national emissions from the petroleum production process, but also consider desisting from the commencement of new petroleum activities. There has been some public debate in Norway on the need to phase out petroleum activities in order to meet climate commitments, but this is not reflected in official government policy. This policy emphasizes the need to meet climate commitments by focusing on reducing climate gas emissions from the sector; but at the same time its goal is to preserve a large petroleum sector through increased recovery from existing fields, development of new discoveries and discovery of new resources.[93]

4.8.2   The General Environmental Regulatory Framework

The main principle is set by the Norwegian Constitution of 1814 (*Grunnlov*), which since 1992 has contained an article on environmental protection.[94] After the revision of the Constitution in 2014, a slightly amended version of that provision is now contained in Article 112, which sets out, *inter alia*, the right of everyone to a healthy natural environment, citizens' right to knowledge about the natural environment and the government's duty to adopt measures to safeguard these principles.[95] Furthermore, Section 10–1 of the Petroleum Act

---

[93]   See further White Paper *Meld St* 28 (2010–11).

[94]   For background information, see HC Bugge, *Environmental Law in Norway* (Wolters Kluwer, 2011), p31.

[95]   The scope of this provision is being considered in a pending climate litigation case. The plaintiffs, Greenpeace and Nature and Youth (*Natur og ungdom*), have challenged the government's decision (formally taken by the MPE) to grant exploration and production licences under the 23rd licensing round in the Barents Sea in an area close

provides that petroleum activities shall be conducted in a prudent manner and shall take due account of the environment, among other things. Further implementing provisions are contained in the Petroleum Regulations.

The main piece of pollution control legislation is the Pollution Control Act. It addresses all types of emissions to air, water (including the sea) and soil, and is therefore applicable to offshore petroleum activities. The Act is based on the principle of prevention and introduces into Norwegian law some central principles of environmental protection, such as the 'polluter pays' principle, the precautionary principle and the substitution principle. The main principle is that planned discharges and emissions are legal only if the operator has applied for and received a permit.

In terms of the protection of nature, the sector-overarching Nature Diversity Act is only partially applicable to activities on the NCS.[96] This approach differs from that chosen for the Norwegian onshore electricity sector, to which the Act applies in general. The Act aims to 'protect biological, geological and landscape diversity and ecological processes through conservation and sustainable use . . ., now and in the future'. It defines a series of principles and obligations which are reflected in the different impact assessment procedures referred to above. Those include, among others: management objectives for habitat types and ecosystems (Section 4); knowledge-based decisions (Section 8); the precautionary principle (Section 9); and ecosystem approach and cumulative environmental effects (Section 10). The political decision to allow only partial application of the key Norwegian nature preservation Act to the petroleum sector may be criticized from an environmental perspective.

Specific obligations in terms of emissions and discharges from petroleum activities are set out in the Carbon Dioxide Tax Act and the Greenhouse Gas Emissions Trading Act. The regulatory framework for HSE also contains provisions relevant to the control of emissions, but focuses on risk and performance obligations of the employers or the operator following a system of internal control which mirrors the obligations under, for example, the Pollution Control Act.

---

to the Ice Edge. This award was based on the decision to move the Ice Edge further north and to open to new areas in 2013 in the southeast of the Barents Sea. The plaintiffs submitted that the award decision is invalid on the ground that it breaches Article 112 of the Constitution, as well as Norway's international commitments under the Paris Agreement to the UNFCCC and other environmental treaties. The Oslo City Court ruled that the grant of exploration and production licences under the 23rd licensing round did not contravene Section 112 of the Constitution and that the award is therefore valid. The decision is now under appeal before the *Borgarting* Court of Appeal.

[96]   Section 2 of Act 100 of 19 June 2009.

Norway is bound by its obligations under international and EEA law, which both contain important provisions in terms of environmental protection. At international level, instruments include the UN Law of the Sea Convention, the Convention on Biological Diversity, the UNFCCC, the OSPAR Convention and the 1979 Geneva Convention on Long-range Transboundary Air Pollution, with its 1999 Gothenburg Protocol to Abate Acidification, Eutrophication and Ground-level Ozone. A large part of the EU legislation on environmental protection has been incorporated into the EEA Agreement (Annex XX to the EEA Agreement) and must therefore be implemented in Norway.

Norwegian legislation includes a large set of regulatory instruments for the purpose of emissions control. Regulatory action is mainly based on regulatory obligations, discharge/emissions permits, compliance monitoring and reporting obligations.[97] In parallel, the industry has progressively developed best practice guidelines, at both national and international level, often in close cooperation with the authorities.[98]

The follow-up of regulatory obligations is performed through compliance monitoring by the agencies with competence in the matter (ie, the NPD and the Norwegian Environment Agency).

The main emissions control obligations applicable to the petroleum industry operating on the NCS are outlined below, classified by air emissions type.

4.8.2.1 Air emissions

Emissions from offshore oil and gas operations represent an important portion of Norway's total emissions to air, and consist primarily of gases containing carbon dioxide (CO_2), nitrogen oxide (NOx), sulphur oxide, methane and non-methane volatile organic compounds (NMVOCs).[99] In 2016 about 29 per

---

[97] The licensees on the NCS are required to report all emissions and discharge data into a dedicated database commonly called the EPIM Environment Hub, which is a joint database for the industry (Norwegian Oil and Gas) and the responsible agencies and directorates. The reporting obligation applies both to planned and approved operational emissions/discharges and to those which occur accidentally. It applies to all fields with production facilities on the NCS. However, emissions/discharges from the construction and installation phase, maritime support services and helicopter traffic are excluded. Reporting is done on an annual basis.

[98] Particular attention should be paid to the *Guidelines* for offshore environmental monitoring, which have been elaborated by the Environmental Agency in collaboration with representatives from the oil and gas industry. The guidelines aim to support companies operating on the NCS in complying with environmental reporting obligations, offering a standardized and comparable reporting frame.

[99] The emissions primarily come from the combustion of natural gas or diesel in turbines, engines and boilers for the purpose of power generation; gas flaring; and the combustion of oil and gas in connection with well testing and well maintenance. Emissions from leaks, gas venting and evaporation from offshore storage and loading

cent of the country's total NOx emissions were generated by the petroleum sector, which also counted for 28 per cent of GHG emissions and about 25 per cent of NMVOC emissions.[100]

At the international level, NOx emissions are regulated by the 1999 Gothenburg Protocol to Abate Acidification, Eutrophication and Ground-level Ozone,[101] to which Norway is a party. The Protocol sets emission ceilings for four pollutants: sulphur, NOx, VOCs and ammonia. The Protocol sets tight limit values for specific emission sources and requires the use of best available techniques to keep emissions down.[102]

At national level, NOx emissions relating to the operation of offshore facilities[103] are regulated by conditions in the PDOs and the approvals of plans for installations pursuant to Section 4–3 of the Petroleum Act. Emissions permits pursuant to the Pollution Control Act are also required for NOx emissions.

NOx emissions have been subject to a NOx tax since 2007. The purpose of the tax is to pursue a cost-effective reduction of NOx emissions. Together with other compliance instruments, it contributes to Norway's compliance with the Gothenburg Protocol. The NOx tax covers emissions from energy generation (eg, turbines and flaring), including from offshore installations. In 2018 the tax rate was 21.94 per kilo of emitted NOx.

A particular form of commitment has been made by Norway's industry associations in the form of an agreement with the government on the reduction of NOx emissions. Fifteen business organizations – including those representing the oil and gas operators on the NCS – have committed themselves to a third period (2018–25) under the NOx Agreement signed with the Ministry of Climate and the Environment.[104] Under this agreement, companies have com-

---

and transport of crude oil can also lead to emissions of hydrocarbon gases (methane and NMVOCs). For an overview of the emissions sources, see www.norskpetroleum.no/en/environment-and-technology/emissions-to-air/, last accessed 2 April 2019.

[100] Source: Environment Web.

[101] Protocol to the Convention on Long-range Transboundary Air Pollution, as revised.

[102] Under the Gothenburg Protocol, Norway has undertaken to reduce its overall NOx emissions by 23 per cent by 2020 compared with the 2005 level.

[103] NOx emissions originated primarily from gas turbines on offshore installations and engines, far ahead of flaring, boilers and well testing. Source: Environment Web.

[104] Environmental Agreement concerning Reduction of NOx Emissions for the Period 2018–25 ('the NOx Agreement 2018–25'), signed on 24 May 2017. The agreement was approved by the European Free Trade Association Surveillance Authority (ESA) on 22 February 2018 (Decision 027/18/COL) as being compatible with Article 61(1) on state aid in the EEA Agreement.

mitted to report emissions and pay a corresponding contribution to the NOx Fund, which is managed by the Confederation of Norwegian Enterprise.[105]

Most NMVOC emissions from offshore oil and gas operations come from storage and loading of crude oil operations offshore (around 50 per cent). NMVOC emissions have considerably decreased during the last decade: they had been reduced by more than 87 per cent in 2013 compared to 2001 figures, and were reduced again by 11 per cent between 2015 and 2016. Such reductions have primarily resulted from the introduction of emissions-reduction technologies in storage ships and shuttle tankers. This is notably based on the requirement to use best available techniques, as provided in the Gothenburg Protocol.[106]

4.8.2.2 GHG emissions and the climate change regime[107]

Methane and CO_2 account for most of the GHG emissions from the NCS. Total methane emissions are continually falling.[108] By contrast, CO_2 emissions remain relatively unchanged, despite the adoption of various policy measures.[109]

Norway's climate policy is primarily driven by its commitments under the UNFCCC and the Kyoto Protocol, as well as under the EEA Agreement.[110] The

---

[105] The NOx Fund was established in 2008 in order to contribute to the development of cost-effective NOx emissions reduction measures, such as innovative technology solutions; www.nho.no/Prosjekter-og-programmer/NOx-fondet/, last accessed 2 April 2019. As a result of the NOx Agreement, companies with activities subject to the NOx tax may choose to contribute to the NOx Fund instead and are thus eligible for the NOx tax exemption scheme. In 2018, the contribution to the NOx Fund for companies in the petroleum sector was fixed at NOK 12 per kilogram of NOx.

[106] Under the Gothenburg Protocol, Norway has committed to reduce its overall NMVOC emissions by 40 per cent by 2020 compared with 2005 levels.

[107] This section builds on the contribution written by author Catherine Banet to the Norway chapter in M Roggenkamp, C Redgwell, I del Guayo and A Rønne (eds) *Energy Law in Europe* (OUP, 3rd edition, 2015).

[108] The offshore petroleum sector accounts for around 10 per cent of these emissions.

[109] Norway's largest source of CO_2 emissions comes from petroleum activities (around 27 per cent), followed by road transport (22.7 per cent), industrial processes (16.3 per cent) and stationary combustion (15.4 per cent); see 'Greenhouse Gas Emissions 1990–2012, National Inventory Report', Report M-137–2014, Norwegian Environment Agency, Chapter 2. The large contribution from the petroleum sector to CO_2 emissions is also explained by the particular shape of the Norwegian economy and the country's energy profile, where hydropower accounts for approximately 96 per cent of onshore power generation.

[110] Norway ratified the UNFCCC on 9 July 1993 and the Kyoto Protocol on 30 May 2002, and became party to the Protocol when the latter entered into force on 16 February 2005. Norway's allotted total annual quota volume (assigned amount) under the Kyoto Protocol's first commitment period (2008–12) is 1 per cent above its emis-

objectives and principles of current national climate policy have developed over time, taking the 2008 political agreement in Parliament as a basis.[111] New climate goals were announced in 2015, ahead of the Paris UNFCCC Conference of the Parties (UNFCCC COP 21/CMP 11).[112] Norway aims to reduce GHG emissions by at least 40 per cent by 2030 compared with 1990 levels, which is aligned with the EU goals announced in 2014.[113] A White Paper on a New Norwegian Commitment for the Period After 2020 was adopted by Parliament on 3 May 2018 and represents Norway's current climate strategy.[114] A Climate Law was adopted on 16 June 2017, establishing emissions reduction targets for 2030 and 2050. The law came into effect on 1 January 2018 and aims for Norway to achieve carbon neutrality by 2050, in quantitative terms – defined as a reduction in greenhouse gas emissions by 80–95 per cent below 1990 emission levels.

The Norwegian government has traditionally favoured the use of general mitigation policy tools based on the polluter pays principle. Norwegian climate policy is based on two cross-sector economic instruments: the CO_2 tax (since

---

sion level in 1990. Norway's emissions exceeded this allotted volume, which resulted in the purchase of allowances to fulfil the commitment. Meanwhile, the government has estimated that it will go beyond the commitment for 2008–12 by 6.6 million tonnes annually, due to the sufficient amount of units it retains in its registry. Norway's commitment under the Kyoto Protocol for the second commitment period (2013–20) was agreed in December 2012 and is that average annual emissions of GHG gases shall be limited to 84 per cent of emissions in 1990. This is an ambitious commitment, which will probably require major cuts in the offshore petroleum and transport sectors. Insufficient emissions cuts will ultimately have to be balanced by the purchase of UN credits – a scenario already foreseen in the national state budget.

[111] Recommendation 145 (2007–08). This agreement was followed in 2010 by the strategy document *Climate Cure 2020*, 15 February 2010, and in 2012 by a climate policy which presented a thorough cross-sector analysis of tools and measures to reduce emissions in Norway – see Report 21 (2011–12) to Parliament, White Paper on Norwegian climate policy. The Climate Cure document contained a specific report on the petroleum sector, which looked at emissions reduction in three areas: energy efficiency, electrification and carbon capture and storage; see summary of conclusions in *Fact Sheet – Options for reduced greenhouse gas emissions in the petroleum sector*, Climate Cure 2020.

[112] 'A new and more ambitious climate policy for Norway', Press Release 28/2015, Office of the Prime Minister, 4 February 2015. See also 'White Paper on new emissions reduction commitments for 2030 – a joint solution with the EU' (*Meld St* 13 (2014–15), *Ny utslippsforpliktelse for 2030 – en felles løsning med EU*), 6 April 2015. Norway signed and ratified the Paris Agreement on 20 June 2016.

[113] It is now envisaged that Norway will enter into a joint agreement with the EU for joint fulfilment of the targets, relying on the EU's climate measures.

[114] *Meld St* 41 (2016–17), *Innst* 253 S (2017–18).

1991) and an emissions trading scheme (ETS) (since 2005). Both apply to the petroleum sector.[115]

The application of the CO_2 tax to the petroleum sector is regulated in Act 72 of 21 December 1990 relating to tax on discharge of CO_2 in petroleum activities on the NCS. Like the NOx tax, the CO_2 tax is intended as a cost-effective measure to reduce emissions. It applies to burnt petroleum and natural gas discharged to air, and to CO_2 separated from petroleum and discharged to air, all emitting from installations used in connection with the production or transportation of petroleum.[116] The Greenhouse Gas Emission Trading Act (Act 99 of 17 December 2004) provides for an emissions allowance and trading scheme, incorporating the EU Emissions Trading Directive 2003/87/EC into Norwegian law.[117] Moreover, gas flaring is prohibited under the Petroleum Act and is allowed only for safety reasons during operation and in connection with certain operational problems. Finally, the electrification of platforms from land-based (renewable) electricity supply and the use of more energy-efficient technologies for power generation are also introduced as measures to reduce emissions from offshore operations.[118]

---

[115] For an analysis of the applicable CO_2 tax and the ETS in Norway, see C Banet, 'Effectiveness in Climate Regulation: Simultaneous Application of a Carbon Tax and an Emissions Trading Scheme to the Offshore Petroleum Sector in Norway', *Carbon and Climate Law Review* (2017) 11(1), ss 25–38.

[116] Section 2 of the Act. Some sectors or products benefit from a reduced rate in accordance with the Energy Tax Directive 2003/96/EC, in order not to alter the overall CO_2 pricing in the affected sectors. Meanwhile, based on the 2012 political agreement, the CO_2 tax for petroleum activities has been considerably increased from NOK 200 per tonne of CO_2 to NOK 410 per tonne, with effect from 1 January 2013.

[117] After a testing phase from 2005, Norway joined progressively the European Union ETS in 2008 and is harmonizing its ETS legislation with that of the EU on an ongoing basis. The Environment Agency is in charge of supervising the implementation of emissions reduction measures and is the reporting authority for the ETS.

[118] Licensees must provide information on the energy solutions proposed for the development of new fields as part of the PDO and the PIO to be submitted to the MPE for approval. Ormen Lange, Troll A, Gjøa and Valhall are already supplied with electricity from land. Some other platforms are only partially powered from shore, due to safety and environmental reasons (eg, Goliat). Electrification has also been approved as a valid supply solution for a series of forthcoming fields: the Martin Linge field in the Northern North Sea and the Edvard Grieg, Ivar Aasen and Gina Krog fields on the Utsira High. The PDO submitted in 2015 for the giant Sverdrup field also includes the development of power from shore as a supply solution.

4.9   CONCLUSION

Ownership of Norwegian subsea petroleum deposits is vested in the State. The Norwegian model for the management of these resources is based on a step-by-step licensing regime that has as its overall aim the promotion of petroleum management through a long-term perspective for the benefit of Norwegian society as a whole.

The licensing regime is governed by the Petroleum Act and appurtenant regulations. The different phases of this regime start with the opening of areas followed by the award of exploration and production licences on the basis of licensing rounds. Successful licensees must enter into JOAs on general terms decided by the State. The exploration and production licence is the most fundamental licence enshrined in the Petroleum Act, as it entails an exclusive right to exploration, exploration drilling and production of petroleum deposits within the area covered by the licence. The development of the field as such is then subject to the approval of a PDO and, in the case of large gas transportation infrastructure, in many cases also a PIO. Production permits and, finally, cessation are also governed by this system. This licensing system has arguably proved a flexible and successful tool in achieving the overall resource management aims, and is well adapted to the specific Norwegian context and requirements.

Upstream oil and gas activities have an impact on their surrounding environment at all stages of resource management. Consequently, the regulatory system seeks to balance the different economic, social and environmental interests relating to the use of relevant sea areas at the stage of opening areas for petroleum exploration; while emissions from petroleum activities are governed through a separate set of environmental regulatory tools.

Overall, the regulatory model enshrined in the Petroleum Act and appurtenant regulations has proved to be a well-functioning and efficient tool to promote the main goals of Norwegian petroleum policy. Some of the regulatory techniques applied may perhaps also serve as inspiration for the regulation of natural resource exploitation elsewhere; while other aspects must be considered on the basis of the specific Norwegian context and may be less suitable for replication.

For regulatory systems where a licensing approach is applied, there is much merit to the stepwise and chronological approach to licensing applied in the Norwegian Petroleum Act. This system recognizes the industry's need for investor certainty by ensuring sufficient predictability throughout the petroleum exploration lifecycle, while at the same time leaving the government discretion to take the necessary decisions at the time those decisions should be made. The most important example in this respect is arguably the relationship between the exploration and production licence and the approval of the PDO, where the former licence provides the licensee with an exclusive exploration right, but where the government nevertheless has the competence to consider and approve the choice of development solution at a later stage pursuant to the latter licence.

Moreover, the regulatory system successfully strikes a balance by setting out the overall principles, competencies and requirements in the Petroleum Act, while leaving the detailed provisions to the more general Petroleum Regulations and, in turn, yet more detailed provisions to subordinate regulations. This has resulted in an easily accessible and transparent regulatory system. The strong emphasis on environmental and climate standards for exploration and production activities is also important to take into consideration for other resource management regimes.

On the other hand, the broad margins of discretion left to the authorities in licensing decisions must primarily be seen as a result of Norwegian policy and regulatory traditions which cannot be viewed in isolation from general principles of Norwegian administrative law. Whether similar margins of discretion should be replicated in other jurisdictions can be fully evaluated only on the basis of the merits of the factual, policy and legal backgrounds of the systems in question.

PART III

United Kingdom

5. Background: United Kingdom

Raphael Heffron, Mohammad Hazrati, Greg Gordon and Darren McCauley

5.1 INTRODUCTION

The main objective of this chapter is to examine the societal context in which the UK system for managing its petroleum resources developed. This chapter will briefly describe British society in terms of its economy, resources and politics.

Although the first oil discovery in the UK dates back to the early 1850s, when shale oil was discovered in Scotland, it took several decades for the country to begin conventional oil production (the first discovery of conventional oil was made in 1919).[1] The conversion of Royal Navy ships from coal to oil and the start of the First World War in 1914 considerably increased domestic demand for oil, which shale was unable to meet. During the Second World War, the search for petroleum was stepped up; but until the 1970s, when rich oil and gas reserves were discovered in the North Sea, the amount of domestic oil and gas production was insignificant. For example, total crude oil production in 1962 including shale, was about 129 000 tons – a minimal amount in comparison to the 12 169 million tons produced in 1976 or the 127 600 million tons produced in 1985.[2]

The discovery of huge amounts of gas in Groningen in Holland in 1959[3] encouraged the UK government to consider the North Sea as a potentially vast resource of petroleum. The government passed the Continental Shelf Act in

---

[1]   Alex Kemp, *The Official History of North Sea Oil and Gas: The Growing Dominance of the State* vol 1 (Routledge 2012) 7 & 8.

[2]   Department for Business, Energy and Industrial Strategy, *Crude Oil and Petroleum: Production, Imports and Exports 1890 to 2016,* www.gov.uk/government/statistical-data-sets/crude-oil-and-petroleum-production-imports-and-exports-1890-to-2011, last accessed 14 October 2017.

[3]   Colin Robinson and Eileen Marshall, *Oil's Contribution to UK Self-Sufficiency* (Heinemann Educational Books 1984) 2.

1964[4] and held the first licensing round that same year, which ultimately led to the first commercial gas discovery in the West Sole field in 1965[5] and subsequently the first gas production in 1967 in the same field. Three years later, in 1970, came the first oil discovery in the Forties Field.[6] Thus, we confine the scope of this chapter to the period between the 1970s and the present, because it is only during this period that oil and gas became significant for the UK government. However, in some sections we examine a wider time period in order to understand evolving trends.

5.2   ECONOMY AND INDUSTRIES

5.2.1   Overall Economic Position and Growth

As Figure 5.1 shows, by the 1970s – the decade in which significant oil and gas resources were discovered – the British economy had experienced three main trends. The first trend began with the Industrial Revolution at the close of the eighteenth century, which transformed Britain's economy and society and made it the richest country in the world.[7] In fact, modern world economic growth began in Britain with the Industrial Revolution,[8] and from this time (1760) until the end of the nineteenth century, the country had the highest gross domestic product (GDP) per capita in the world.[9] By using new technologies in several industries, such as textiles, iron and coal, reducing prices and increasing exports, Britain gained a dominant economic position.[10]

The second trend is from the late nineteenth to early twentieth centuries, when Britain lost its dominance in the global economy. The industrial output of Germany and the US, especially in the heavy industries, gradually surpassed

---

[4]   Kemp (n 1) 34.
[5]   Fred Atkinson and Stephen Hall, *Oil and The British Economy* vol 6 (Routledge 2016) 18.
[6]   Kemp (n 1) 236.
[7]   Justin Yifu Lin, 'The Latecomer Advantages and Disadvantages: A New Structural Economic Perspective' in Martin Andersson and Tobias Axelsson, *Diverse Development Paths and Structural Transformation in the Escape from Poverty* (Oxford Scholarship Online 2016) 43 and 44.
[8]   Jeffrey Sachs, *The Age of Sustainable Development* (Columbia University Press 2015) 74.
[9]   Knick Harley, 'The Legacy of Early Start' in Roderick Floud, Jane Humphries and Paul Johnson (eds), *The Cambridge Economic History of Modern Britain: 1870 to Present* vol 2 (CUP 2014) 6.
[10]   Ibid pp 2–11.

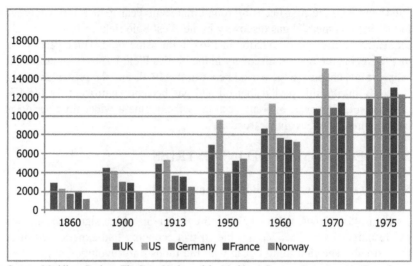

Source: Maddison Project, *The First Update of the Maddison Project, Re-estimating Growth Before 1820*, available at www.ggdc.net/maddison/maddison-project/abstract.htm?id=4, last accessed 14 October 2017.

Figure 5.1   UK GDP per capita 1990 in GK$

that of Britain, and the US became the leading industrial power.[11] The outbreak of the First World War in 1914 disrupted commerce and finance and caused significant debt and inflation in European countries, which in turn exacerbated the gap between the US and its economic rivals. In this new era, the US achieved a superior position in the global economy; but in general, the UK still enjoyed better economic performance than countries such as France, Norway and West Germany during the 1950s. This trend changed in the period from 1950–73 (Europe's Golden Age),[12] when the UK began to lag behind several European countries. Thus, by around 1960–70, when commercial discovery occurred, the UK economy was one of the most developed in the world, although behind the US; while countries such as Germany and France were rapidly narrowing the economic gap with the UK.

---

[11]   Nicholas Crafts, 'Economic Growth During the Long Twentieth Century' in Roderick Floud, Jane Humphries and Paul Johnson (eds), *The Cambridge Economic History of Modern Britain: 1870 to the Present* vol 2 (CUP 2014) 27.

[12]   The 'Golden Age' of European economic growth refers to the period between the Second World War and 1973, during which many European countries (except Britain) experienced rapid economic growth and reduced the large productivity gap with the United States.

However, the UK's real GDP per capita from 1967–2016 increased by more than 100 per cent, from £11 474 in 1967 to £28 982 in 2016. Undoubtedly, the UK's oil and gas industry has played a significant role in this achievement.[13]

5.2.2 Shipping and Shipbuilding

The geographical position of the UK, as an island nation, has inevitably made shipping a vital sector for the country. Over the centuries, travel, trade and even war and defence were possible only via sea and through the use of ships.[14] However, the Industrial Revolution in the late eighteenth century galvanized Britain's shipping industry. Between the 1870s and the 1940s, the British ship-building industry held a dominant position on the market, although its share of the market consistently decreased from around 80 per cent in the 1890s to 60 per cent in 1914.[15] Britain's dominant position in shipping was largely because of its access to a large market which offered opportunities for and access to mass production, specialization, cheap supplies of raw materials and skilled men.[16] In 1920 the industry still directly employed over 300 000 people and consumed 30 per cent of pig iron – figures which demonstrate its huge size at the time.[17] During the two World Wars, the British shipbuilding industry lost its dominance in the global market and its share of the market fell to less than 40 per cent. This trend worsened in the recession of 1958–61, when in just three years employment and merchant tonnage under construction decreased by 43 per cent and 45 per cent, respectively. After many years of being the world leader, the British shipping industry had lost its position.[18]

Despite this disappointing failure in the global market, the industry still con-tributes to the UK economy today through gross value added (GVA), business turnover and job creation. It is estimated that in 2015, the industry contributed £4.3 billion in GVA and £13.9 billion in business turnover, and created 152

---

[13]   Bank of England, *A Millennium of Macro Economic Data for the UK: Version 3.1, Finalized 2017,* available at www.bankofengland.co.uk/research/Pages/datasets/default.aspx, last accessed 14 October 2017.

[14]   Mark Dunkley and Paul Stamper, 'Ships and Boats: Prehistory to 1840' (Historic England, 8 July 2016), https://historicengland.org.uk/images-books/publications/iha-ships-boats, last accessed 14 October 2017.

[15]   Alastair J Reid, *The Tide of Democracy: Shipyard Workers and Social Relations in Britain: 1870–1950* (MUP 2010) 22–24.

[16]   Sidney Pollard, 'British and World Shipbuilding, 1890–1914: A study in Comparative Costs' (1957) 17 *The Journal of Economic History,* 426, 444.

[17]   Ibid, p444.

[18]   Lewis Johnman and Hugh Murphy, *British Shipbuilding and the State Since 1918: A Political Economy of Decline* (University of Exeter Press 2002) 130.

600 jobs (more than 35 per cent of which were filled by British employees).[19] The international transport of freight is the largest constituent activity within the shipping industry, contributing £2.7 billion in GVA and supporting 29 800 jobs in 2015.[20]

The shipping industry (including the transportation of passengers and freight in both inland and international waters) is usually examined within the broader maritime sector, which includes the port industry (eg, warehousing and storage, port activities and management, stevedores and passenger handling and border agencies), the marine industry (eg, shipbuilding, boatbuilding, marine renewable energy and marine support activities for offshore oil and gas activities), and the maritime business services industry (eg, insurance, legal services, shipbroking, education and accountancy).[21] The marine industry (the part in which shipbuilding is placed) is the largest industry in the maritime sector. It contributed an estimated £6.45 billion in GVA, supported 99 500 jobs and created £17.9 billion in business turnover, once indirect and induced economic channels are taken into account, in 2015. Of the different kinds of activities in the marine industry, marine oil and gas support had the largest aggregate GVA impact in 2015, with a £4.3 billion contribution (including indirect and induced impact) to GDP.[22]

Access to Europe's single market has offered significant benefits for the UK shipping industry, as much of the UK's international trade with EU members has been via sea. The post-Brexit era will pose challenges for the industry; for instance, UK flagged ships might lose their right to operate in the domestic trades of EU member states which maintain flag-based cabotage restrictions.[23]

---

[19] Centre for Economics and Business Research (CEBR), *The Economic Contribution of the UK Marine Industry: A Report for Maritime UK 2017,* available at www.maritimeuk.org/documents/188/Cebr_Maritime_UK_Marine_finalised.pdf, last accessed 14 October 2017.

[20] Ibid.

[21] Oxford Economics, *The Economic Impact of the UK Maritime Services Sector: Shipping.*

[22] Ibid.

[23] House of Commons Library, *Future of the UK maritime industry,* available at http://researchbriefings.parliament.uk/ResearchBriefing/Summary/CDP-2017-0009, last accessed 15 October 2017.

5.2.3 Fisheries

Fishing is not considered to be one of the UK's top industries in terms of economic output.[24] Less than 0.1 per cent of UK GDP is from fisheries.[25] The UK is known as a net importer of fish, with around 238 000 tons imported in 2015, worth about £1.3 billion. Before the First World War, the UK tonnage of fish landed reached a peak of 1.2 million tons in 1913.[26] From the mid-1970s onwards, there has been a steady decline in the number of UK vessels and tonnage landed; for example, in 1996 there were 8667 UK fishing vessels, which landed about 892.3 thousand tons (in the UK and abroad); while just six years later, in 2002, this amount had declined by more than 12 per cent in terms of numbers of vessels, falling to 7570 vessels, while the amount of fish landed had decreased by more than 23 per cent to 685.5 thousand tons. This trend continued until 2015, when the number of vessels fell to 6187, but tonnage landed increased slightly to 708 thousand tons. Further, the number of UK fishermen over the last century has considerably reduced, from 47 824 in 1938 to 28 254 and 12 107 in 1960 and 2015, respectively.[27]

5.2.4 Services

The service industry dominates the UK economy, accounting for around 80 per cent of GDP in 2015, while the manufacturing, construction and agriculture industries accounted for just 10, 6 and 1 per cent, respectively.[28] The service industry also has the largest share of UK employment, at 85 per cent, accounting for 23.8 million employees. The different industries have not always had the same weight in the UK economy. In 1948 the service industry accounted

---

[24]   The Marine Socio-Economic Project (MSEP), *How important is fishing to the UK economy?*, available at www.mseproject.net/data-sources/doc_download/122-8 -fishing-and-uk, last accessed 14 October 2017.

[25]   Department for Environment, Food and Rural Affairs, *The Government has announced it will withdraw from the London Fisheries Convention*, available at www .gov.uk/government/news/uk-takes-key-step-towards-fair-new-fishing-policy-after -brexit, last accessed 15 October 2017.

[26]   House of Commons Library, *UK Sea Fisheries Statistics: November 2016*, available at researchbriefings.files.parliament.uk/documents/SN02788/SN02788.pdf, last accessed 15 October 2017.

[27]   Ibid.

[28]   House of Commons Library, *Industries in the UK*, available at http:// researchbriefings.parliament.uk/ResearchBriefing/Summary/SN06623, last accessed 16 October 2017.

for 46 per cent of GDP, while production industries accounted for 42 per cent of GDP.[29]

The service industry is comprised of the retail sector, the financial sector, the public sector, business administration, leisure and cultural activities. The main components of the industry include:

- distribution, hotels and restaurants;
- transport, storage and communication;
- business services and finance; and
- government and other services.

In 2016 the proportion of UK services that were exported was 44 per cent, which is considered one of the highest shares of any advanced economy; this is estimated to reach 50 per cent by 2026. In 2016 machinery and electrical products and financial services were ranked as the first and second exported products respectively; but by 2026 it is expected that their rankings will be reversed and financial services will become the UK's primary export industry with machinery and electrical products second.[30]

5.2.5   The Petroleum Sector's Impact on the UK Economy

The UK is the second largest oil producer in Organisation for Economic Co-operation and Development (OECD) Europe, after Norway, and the third largest gas producer after Norway and the Netherlands.[31] Since the first commercial discovery in the North Sea in 1967, more than 43 billion barrels of oil and gas have been recovered, contributing almost £330 billion in corporate tax to the Treasury.[32] Although this figure has been falling in recent years, it is still a substantial contributor to the UK economy, energy security and jobs. In 2016 alone, the industry added about £17 million to the UK's balance of trade

---

[29]   Office for National Statistics (ONS), *UK Service Industries: Definition, Classification and Evolution, 2013,* available at
http://webarchive.nationalarchives.gov.uk/20160108050516; www.ons.gov.uk/ons/rel/naa1-rd/national-accounts-articles/uk-service-industries--definition--classification-and-evolution/index.html, last accessed 16 October 2017.

[30]   Barclays Bank, *UK Trade Outlook 2016-2026: What to Expect in the Next Decade for UK Exports, 2016,* available at www.barclayscorporate.com/content/dam/corppublic/corporate/Documents/Operating_Internationally/UK-Trade-Outlook-Report.pdf, last accessed 16 October 2017.

[31]   House of Commons Library, *UK Offshore Oil and Gas Industry,* available at http://researchbriefings.parliament.uk/ResearchBriefing/Summary/CBP-7268, last accessed 17 October 2017.

[32]   Oil and Gas UK, *Economic Report 2016,* available at https://oilandgasuk.co.uk/product/economic-report-2016, last accessed 16 October 2017.

and supported over 300 000 jobs across the country.[33] In addition, the industry provides more than 60 per cent of domestic oil and gas demand.[34] In 2016 oil and gas accounted for more than 75 per cent of the UK energy mix.[35] Most of the UK's reserves are off the coast of Scotland. The complexity and maturity of the UK Continental Shelf (UKCS) and the smaller size of most new discoveries are features of this basin, making operating in this area relatively costly.

However, the industry faces some important challenges, one of which is the level of investment. In 2016 just £8.3 billion was invested in the UKCS and two new fields were approved, compared to £15 billion invested in 2013 and ten projects approved. This shows an urgent need for fresh capital in the industry.[36] In response to this challenge, the government has started to abolish petroleum tax revenue by permanently reducing the rate from 35 per cent to 0 per cent, to simplify the fiscal regime for investors and increase the attractiveness of UKCS projects.[37] In addition, it reduced the supplementary charge from 20 per cent to 10 per cent. These reductions in taxes led to just £43 million in revenue for the Treasury in financial year 2015–16, compared to £2217 billion in 2014–15.[38]

The UK's decision to leave the EU has led to uncertainty for the whole of the macro economy and particularly for investors. In addition, the ability for UK oil and gas industry products and services to access EU markets will remain unclear until the final agreement on the future relationship between the UK and the EU has been concluded.[39] However, in the short term, the weakening of sterling following the referendum has made UK exports more competitive and was a positive benefit for oil and gas producers and supply chain companies.[40] The long-term effects of Brexit will depend on the final

---

[33] Oil and Gas UK, *Economic Report 2017,* available at http://oilandgasuk.co.uk/wp-content/uploads/2017/09/Economic-Report-2017-Oil-Gas-UK.pdf, last accessed 16 October 2017.

[34] Ibid.

[35] Ibid.

[36] Oil and Gas UK, *Business Outlook 2017*, available at http://oilandgasuk.co.uk/wp-content/uploads/2017/03/Business-Outlook-2017-Oil-Gas-UK.pdf, last accessed 17 October 2017.

[37] Her Majesty's Revenue and Customs, *Oil and Gas Taxation: Reduction in Petroleum Revenue Tax and Supplementary Charge*: *16 March 2016*, available at www.gov.uk/government/publications/oil-and-gas-taxation-reduction-in-petroleum-revenue-tax-and-supplementary-charge/oil-and-gas-taxation-reductionin-petroleum-revenue-tax-and-supplementary-charge#policy-objective, last accessed 17 October 2017.

[38] GOV, *Government Revenues from UK Oil and Gas Production*, available at www.gov.uk/government/uploads/system/uploads/attachment_data/file/534505/UKCS_Tax_Table_July_2016.pdf, last accessed 17 October 2017.

[39] Oil and Gas UK, *Economic Report 2017* (n 33).

[40] Ibid.

agreement on the future relationship between the UK and the EU, and on how the cost of goods and services may change. The oil and gas market is capital intensive, so to maintain an investor-friendly environment in the post-Brexit era, investor confidence and access to finance should be secured. The positive influence of EU legislation on environmental standards in the energy industry and EU energy policy security must be compensated for post-Brexit.[41]

The steep decline of the oil price since 2014, with the average price of oil (Brent) falling from over US$99 per barrel (bbl) to an average US$43.5 bbl by 2016, has been another challenge for the industry.[42] Of course, the volatility of the oil price is a characteristic feature of the oil industry and is not limited to this period. For example, following the Iraqi invasion of Kuwait, the oil price sharply increased in just one month from US$17.80 bbl in July 1990 to US$31.80 in August 1990.[43]

In 2015, to maximize economic recovery in the onshore and offshore oil and gas industry, including the UKCS, the UK government established a new body called the Oil and Gas Authority (OGA).[44] This has the objectives of reducing operating costs and improving efficiencies.[45] The available data suggests that the initiatives proposed to achieve these aims have been successful. In 2014 the unit operating cost per barrel was about US$29.30, which was success-fully reduced to US$16 bbl in 2016.[46] During the same timeframe there was an increase from 65 per cent to 75 per cent in production efficiency which, together with new field start-ups, has led to an increase in production of about 16 per cent. The long-term vision for the UK oil and gas industry (Vision 2035) sets out both the potential opportunities for the industry and how it can be secured. Vision 2035 has two key objectives: increasing the productive life of the UKCS and doubling the long-term opportunities for the supply chain.[47]

---

[41]   House of Commons, *Brexit Impacts Across Policy Areas, 2016*, available at   researchbriefings.files.parliament.uk/documents/CBP-7213/CBP-7213.pdf,   last accessed 19 October 2017.

[42]   Statista, *Average Annual Brent Crude Oil Price From 1976 to 2016*, available at www.statista.com/statistics/262860/uk-brent-crude-oil-price-changes-since-1976/, last accessed 18 October 2017.

[43]   House of Commons, *Oil Price*, available at http://researchbriefings.parliament .uk/ResearchBriefing/Summary/SN02106, last accessed 18 October 2017.

[44]   Department of Energy and Climate Change, *Government Response to Sir Ian Wood's UKCS: Maximizing Economic Recovery Review, July 2014*, available at www .ogauthority.co.uk/media/1018/wood_review_government_response.pdf, last accessed 18 October 2017.

[45]   Ibid.

[46]   Oil and Gas UK, *Economic Report 2016* (n33).

[47]   Oil and Gas Authority, *Vision 2035*, available at www.ogauthority.co.uk/media/ 3196/vision-2035-overview-january-2017.pdf, last accessed 18 October 2017; and Oil

5.3   EMPLOYMENT AND HUMAN RESOURCES

As Table 5.1 shows, the employment rate over a 50-year period has increased by less than 3 per cent; these statistics are from the Office of National Statistics (ONS). The shares of the different sectors of the UK economy – in particular, those supporting and creating jobs – have significantly changed. The manufacturing sector employed 32 per cent of all employees in 1971, but had lost most of this share and decreased to only 8 per cent by 2016. This trend is more surprising when a longer timeframe is considered; for example, in 1841, the manufacturing and agriculture sectors employed 36 per cent and 22 per cent of all employees, respectively.[48]

The largest sector in 1841 in terms of employment was manufacturing, but this gradually declined over the decades, mainly due to the deliberately restrictive economic policy of the UK government between 1980–82.[49] On the other hand, the service sector increased from 55 per cent in 1971 to more than 80 per cent in 2016.

Further, since 1971, the composition of the workforce in terms of male and female employment rates has also changed. The female employment rate increased from 52 per cent to 69 per cent during this period, while the male employment rate declined from 91 per cent to 70 per cent. The growing rate of female employment is a trend that began following the Second World War.[50] There may be several factors for this trend, but the shift from manufacturing to services is important in this regard. Traditionally, more women work in the service sector, while more men work in manufacturing.[51]

Another important point to note is the rate of employment in 2016, which reached 74.5 per cent – the highest rate since 1971. In 2016, there were 1 660 000 people not in work but available and looking for a job, of whom 46 per cent were women and 54 per cent were men.

---

and Gas UK, *Economic Report 2018*, available at https://oilandgasuk.co.uk/product/economic-report-2018/, last accessed 30 October 2018.

[48]   ONS: The National Archives, *170 Years Of Industrial Changes Across England and Wales*, available at http://webarchive.nationalarchives.gov.uk/20160106001413/ http://www.ons.gov.uk/ons/rel/census/2011-census-analysis/170-years-of-industry/ 170-years-of-industrial-changeponent.html, last accessed 18 October 2017.

[49]   Fred Atkinson and Stephen Hall, *Oil and the British Economy* (Palgrave Macmillan 1984) 200.

[50]   Timothy J Hatton, 'Population, Migration and Labour Supply: Great Britain 1871–2011' in Roderick Floud, Jane Humphries and Paul Johnson (eds), *The Cambridge Economic History of Modern Britain: 1870 to the Present* vol 2 (CUP 2014) 117.

[51]   ONS, *Women in Labour Market, 2013*, available at http://webarchive .nationalarchives.gov.uk/20160108012507; http://www.ons.gov.uk/ons/dcp171776 _328352.pdf, last accessed 18 October 2017.

Table 5.1   Employment structure

Year	1971	2016
Population total	55,928,000	65,684,000
Persons in employment, including self-employment (aged 16 to 64) (%)	24,507,000 = 72	31,800,000 = 74,5
Male employment rate (of all working-age men) (%)	91	79
Female employment rate (of all women of working age, 16–62) (%)	52	69
Unemployment rate (%)	4.1	4.9
Share of persons employed, by sector:		
Services (%)	55	83
Manufacturing, mining and quarrying (%)	32	8
Construction (%)	6	6
Agriculture (%)	3.11	1

Source: Bank of England, *A Millennium of Macroeconomic Data*, available at www .bankofengland.co.uk/research/Pages/datasets/default.aspx, last accessed 18 October 2017.

5.3.1   Education

One of the most significant changes to the workforce has been the increase in human capital – in particular, education. Until the mid-nineteenth century, formal education in the UK was limited. However, by 1914 only 1 per cent of brides and grooms could not sign their names.[52] Overall participation in higher education increased from 3.4 per cent in 1950 to 8.4 per cent in 1970, and from 19.3 per cent in 1990 to 33 per cent in 2000. In 1970, 72 per cent of university graduates were men and 27 per cent were women. In addition, in 1972 the minimum school leaving age increased to 16.

In terms of education, there have been many differences between the sexes. From 1970 to the 2000s, the number of male graduates was always higher than the number of female graduates. But by the 2000s, more women received a first degree than men. The percentage of female graduates increased from 53 per cent in 1998 to 58 per cent in 2005. In the 1990s women primarily studied social sciences, education, health, arts and the humanities; while men primarily

---

[52]   Timothy Hatton, 'Population, Migration and Labour Supply: Great Britain 1871–2011' in Roderick Floud, Jane Humphries and Paul Johnson, *The Cambridge Economic History of Modern Britain* (CUP 2014) 111.

studied science and engineering.[53] By 2014–15, the most popular subjects for male undergraduate students were business, engineering and biological sciences; while the most popular subjects for female undergraduate students were allied health subjects, business and biological sciences.[54]

5.3.2 Time at Work or Not at Work

In the UK, young people are expected to seek employment from the age of 16. In 2016 the number of people aged 16–24 in full-time education was around 29 per cent, while 71 per cent were not in full-time education. For the same age group, the male and female employment rate was equal at around 55 per cent.

Maternity leave is 52 weeks; of those 52 weeks, only two weeks (or four weeks if the job is in a factory) after the birth are compulsory. In addition, if an employee's partner has a baby, adopts a child or has a baby through a surrogacy arrangement, the partner may be entitled to one or two weeks' paid paternity leave.

In 1970 the working week was 42 hours,[55] while today it is 37.5 hours per week. People in part-time work, on average, work 16.3 hours per week. Most individuals who work a five-day week must receive at least 28 days' paid annual leave per year; this is the equivalent of 5.6 weeks' holiday. Part-time workers receive less paid holiday than full-time workers. However, there are different rules for people who work at night and on Sundays. Further, there are eight bank holidays; and if a bank holiday (public holiday) falls on the weekend, a substitute weekday becomes a bank holiday – normally the following Monday.

Part-time work is more common for women than men; for example, 41 per cent of women and 12 per cent of men worked part time in 2016.

In the public sector, the highest number of people worked in health and other social work (National Health Service (NHS)) and education. In 2016, 17 per cent of all people in work were employed in the public sector (the lowest

---

[53] Stephan Vincent-Lancrin, 'The Reversal of Gender Inequalities in Higher Education: An On-going Trend' in OECD, *Higher Education to 2030: Demography* (OECD 2008), available at www.oecd.org/education/skills-beyond-school/highereducationto2030vol1demography.htm, last accessed 18 October 2017.

[54] Universities UK, *Higher Education in Facts and Figures, 2016,* available at www.universitiesuk.ac.uk/facts-and-stats/data-and-analysis/Documents/facts-and-figures-2016.pdf, last accessed 18 October 2017.

[55] Michael Huberman and Chris Minns, 'The Times they are not Changing': Days and Hours of Work in Old and New Worlds, 1870–2000' (2007) 44 *Explorations in Economic History* 538.

proportion since comparable records began in 1999), while the remaining 83 per cent worked in the private sector.

One of the traditional concerns about the UK workforce is sickness. However, this trend sharply declined during the period 1993–2016. In 2016 the main reasons for absence were minor illnesses such as coughs and colds (32 per cent). The groups that experienced the highest rate of sickness absence were women, older workers, those with long-term health conditions, smokers and public health sector workers. Another reason for absence from work is labour disputes, which cause many lost work days. The number of work days lost due to labour disputes in 2016 was 322 000 – an increase from 2015, mainly because of a dispute involving junior doctors in the NHS in England (40 per cent total work days lost).

5.4   DEMOGRAPHICS

5.4.1   Population Increase and Fertility

The UK population was 55.6 million in 1970, growing to 65.5 million in 2016[56] – an increase of approximately 10 million or 17.8 per cent. In 2016 the fertility rate was 1.81 children per woman, while in 1970 the fertility rate was 2.40.[57] The average age of mothers has increased to 30.4 years. As shown in Figure 5.2, the UK experienced a baby boom in the 1960s, which has resulted in a larger population in their 40s and 50s today. In contrast, in the 1970s – in particular, 1977 – there was low fertility for two reasons: the Abortion Act of 1967 and a lower population of individuals in their 30s and early teens. In 2012 the UK witnessed another baby boom, with 813 000 births recorded – the highest number since 1990.

In terms of women's age and fertility rate, from 1970–2004, women aged between 25–29 had the highest fertility rates. In 2004 this pattern changed and women aged between 30–34 now have the highest fertility rates. In addition, fewer women under 20 and over 40 now have children. Rising fertility rates at an older age are due to several reasons, including increased female partici-

---

[56]   OECD, *Population (indicator) 2017,* available at www.oecd-ilibrary.org/social-issues-migration-health/population/indicator/english_d434f82b-en, last accessed 13 October 2017.

[57]   OECD, *Fertility rates (indicator) 2017,* available at www.oecd-ilibrary.org/social-issues-migration-health/fertility-rates/indicator/english_8272fb01-en,   last accessed 13 October 2017.

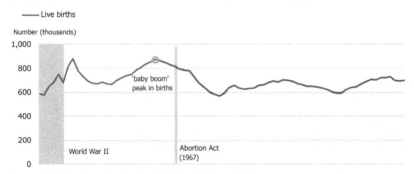

Source: ONS, *Births in England and Wales: 2015*, available at www.ons.gov.uk/peoplepopulationandcommunity/birthsdeathsandmarriages/livebirths/bulletins/birthsummarytablesenglandandwales/2015, last accessed 19 October 2017.

Figure 5.2   Number of live births 1940–2016

pation in higher education and the workforce and the increased importance of a career.[58]

5.4.2   Immigration

In 2016 approximately 6 million people of non-British nationality and 9.2 million people who were born abroad were living in the UK. In the year ending March 2017, net migration was estimated at over 246 000. In 2016, one in seven people living in the country were born outside the UK and one in 11 people were non-British nationals.[59] In 2015 44 per cent of those migrating to the UK were non-EU nationals, 43 per cent were nationals from other EU countries and 13 per cent were British nationals.

Since 1994, the number of people migrating to the UK has been greater than the number of people emigrating from the UK. Conversely, from the 1960s to the 1990s, the number of emigrants was greater than the number of immigrants. However, in the twentieth century, the number of individuals migrating from and to the UK was relatively balanced. From 1991 to 2016, immigration increased to 79 per cent, leading to net immigration. During the

---

[58]   ONS, *Births in England and Wales: 2016*, available at www.ons.gov.uk/peoplepopulationandcommunity/birthsdeathsandmarriages/livebirths/bulletins/birthsummarytablesenglandandwales/2016, last accessed 19 October 2017.

[59]   ONS, *Migration Levels: What Do You Know About Your Area? 2017*, available at https://visual.ons.gov.uk/migration-levels-what-do-you-know-about-your-area/, last accessed 13 October 2017.

first three decades of the twentieth century, the UK experienced net emigration of around 80 000 a year. For the next three decades, from 1931 to 1961, the flow of migration turned inwards, with average net immigration of around 19 000 a year.

In 2015 the EU countries with the largest inflows of foreign nationals were Germany (1 460 000), the UK (548 000) and Spain (290 000). The only migrant group that is larger by nationality than by country of birth comprises migrants from the eight 'accession' countries that joined the EU in May 2004. Among EU nationals, Poland is the most common non-UK country of birth and non-British nationality in the UK. In 2016 the number of Polish nationals residing in the UK reached 1 million.[60]

Between 1997 and 2017, the main reasons for immigrating to the UK were work and study. However, the reasons for immigration differ between EU and non-EU nationalities. The most common reason for EU nationals to move to the UK is work (60 per cent), while study is the most common reason for non-EU nationals to move to the UK. This difference is likely caused by different rights in different countries and the effect of government policies and factors such as economic conditions in origin countries.[61]

The 1990s witnessed a surge in asylum applications to about 50 000 annually.[62] In 2015, 32 733 applications for asylum were filed in the UK – around 5.3 per cent of total immigration in 2015. In 2017 a total of 16 211 people were granted asylum, resettlement or an alternative form of protection – 7 per cent more than in 2015. In 2016 the largest number of applications for asylum came from nationals from Iran (4184), followed by Pakistan (2870) and Iraq (2672).[63]

The UK's migrant population is concentrated in London. Around 38 per cent of people living in London were born outside the UK. After London, the English regions with the highest proportions of their population born abroad are the West Midlands (13.1 per cent), the Southeast (13.1 per cent) and the

---

[60]   ONS, *Population of the UK by country of birth and nationality: 2016*, available   at   www.ons.gov.uk/peoplepopulationandcommunity/populationandmigration/ internationalmigration/bulletins/ukpopulationbycountryofbirthandnationality/2016, last accessed 13 October 2017.

[61]   ONS, *Migration Statistics Quarterly Report: August 2017*, available at www.ons.gov.uk/peoplepopulationandcommunity/populationandmigration/ internationalmigration/bulletins/migrationstatisticsquarterlyreport/august2017,   last accessed 13 October 2017.

[62]   Hatton, (n 53) 103.

[63]   ONS, *Migration Statistics Quarterly Report: August 2017*, available at www.ons.gov.uk/peoplepopulationandcommunity/populationandmigration/ internationalmigration/bulletins/migrationstatisticsquarterlyreport/august2017,   last accessed 13 October 2017.

East (13.0 per cent). Of all regions of the UK, the Northeast has the lowest proportion of its population born abroad (5.8 per cent), followed by Wales (6.0 per cent), Northern Ireland (7.6 per cent) and Scotland (8.6 per cent).

5.4.3 Diversity and Homogeneity

The 1991 population census was the first to ask about individuals' ethnicity. Asking about ethnicity in the 1970s was a challenging question, and was ultimately abandoned. While before the Second World War there were a small number of ethnic groups in Britain, following the Second World War the number of ethnic groups grew due to the 1948 British Nationality Act.

The main minority ethnic groups came from Commonwealth countries and Pakistan. In 1961 approximately 500 000 people from Commonwealth countries and Pakistan were living in the UK. By 1971 the number of minority ethnic individuals had reached 1.3 million. In the 1970s and 1980s the major migrant groups arriving in the UK were Indian and Pakistani, as a result of family reunification.[64]

The 1991 census was the first ever to collect information about ethnicity in the UK. In 1991 the population was nearly 54.9 million, with over 3 million individuals (5.5 per cent) identifying as an ethnic minority. In 2011 one in five people (20 per cent) identified as an ethnic group.

5.4.4 Spatial Population Structure

The UK comprises four countries: England, Scotland, Wales and Northern Ireland. The national government at Westminster represents all four countries. England has the largest population with 84.2 per cent, followed by Scotland (8.2 per cent), Wales (4.7 per cent) and Northern Ireland (2.8 per cent). Historically, there were 91 counties: 39 in England, 33 in Scotland, 13 in Wales and six in Northern Ireland.[65] However, the current structure of local government is different from area to area. For example, England has 353 local authorities, of which there are five different kinds: county councils; district councils; unitary authorities; metropolitan districts; and London boroughs.

---

[64] Pace for Opportunity, *Population of Ethnic Minorities in Great Britain*, available at https://workplace.bitc.org.uk/sites/default/files/paper_1_rfo_business _developement_activities_of_ethnic_minorities_in_gb_paper_1_population_of_ethnic _minorities_in_great_britain.pdf, laT accessed 14 October 2017.

[65] Jonny Muir, *The UK's County Tops* (Cicerone 2011) 13.

In 2016 the UK had the equivalent of 263 people per square kilometre.[66] The UK population is incredibly dense near major cities such as London, Portsmouth and Liverpool, with density of at least 200 people per square kilometre. Outside these areas, density generally varies between 11 and 100 people per square kilometre. In central Wales and most of Scotland, population density is incredibly low, with a density of one to ten people per square kilometre.[67] In 2016 an estimated 2.85 million people moved among local authorities in England and Wales. The two regions with the greatest movement were London and the Southeast, although they also have the largest populations.[68]

London is the UK's capital city and has a population of 8 787 000. The city has played a prominent role in the country's history and was home to the biggest concentration of industries, such as clothing, footwear and furniture. Major infrastructure was built, including the port of London (the country's largest), and the UK also built a vast network of railways and roads emanating from London.[69] The high-growth areas of London show the highest levels of net international migration and negative net internal migration.

The younger demographics of places such as London, Northern Ireland, Manchester and Birmingham tend to lead to higher numbers of births than deaths, resulting in relatively high levels of positive natural change.[70]

Although the UK geographical landscape is predominantly rural, at least 60 per cent of the population live in urban areas. During the last two centuries, the geographical distribution of the UK population changed significantly. Between 1841 and 1911, the rural areas of England and Wales lost around 4.5 million people through migration. The major destinations were the mining cities of Wales, Manchester, Liverpool and, above all, London. Due to the higher wages (1860 to 1913) of urban labour (40 per cent) compared to agricultural labour, people began moving to urban areas. However, this trend changed following the First World War, mainly due to declines in major industries such as textiles, shipbuilding and coal.

---

[66]   ONS, *Overview of the UK population: March 2017*, available at www.ons.gov.uk/peoplepopulationandcommunity/populationandmigration/populationestimates/articles/overviewoftheukpopulation/mar2017, last accessed 14 October 2017.

[67]   Alex Jackson, 'Population, Distribution and Density of the UK' (Geography As Notes), available at https://geographyas.info/population/uk-population-distribution-density/, last accessed 14 October 2017.

[68]   ONS, *Statistical Bulletin: Population Estimates for UK, England and Wales, Scotland and Northern Ireland: mid-2016*, available at www.ons.gov.uk/peoplepopulationandcommunity/populationandmigration/populationestimates/bulletins/annualmidyearpopulationestimates/mid2016, last accessed 14 October 2017.

[69]   PJ Waller, *Town, City and Nation: England 1850–1914* (2nd OUP 1999) 25–30.

[70]   ONS, *Statistical Bulletin: Population Estimates for UK, England and Wales, Scotland and Northern Ireland: mid-2016* (n 69).

Today, migration is inter-urban and mostly due to job opportunities, wages, education and retirement. People who own their own homes and council tenants are less likely to move compared to private renters.

5.5 INFRASTRUCTURE AND ACCESS TO MARKETS

5.5.1 Physical Infrastructure and Transportation

The UK is in general served by a well-developed road network and all but the most remote settlements having a road connection.[71] Causeways and bridges provide road connections even to some island communities,[72] although these cannot always be depended upon in poor weather. Boat services (generally operated by private sector companies under the auspices of public concessions issued after tender) link the more remote islands to more populous neighbouring islands and/or to the mainland. The overwhelming majority of passenger transport kilometres recorded within the UK are for travel by road.[73]

Road standards within the UK are variable. A reasonably extensive motorway network exists, although construction of the system commenced in the late 1950s and motor vehicle usage has increased considerably since then. In some areas the system is congested and in need of modernization. Population density means that the motorway system is generally better developed within England than in the other constituent parts of the UK, although even in England some counties are not connected to a motorway (eg, Norfolk). The highly populous central belt of Scotland is served by a well-developed motorway network.

Away from the motorways, roads are classified into numbered A and B roads and un-numbered classified and unclassified roads. Some A roads are dual carriageways. B roads are often quite narrow and can be winding and undulating. In more remote areas, many un-numbered roads are single-track roads with passing places. In general, main roads seek to bypass major cities in order to ease congestion, although the building of bypasses remains an ongoing

---

[71] In the case of some isolated coastal villages, such as Crovie in Aberdeenshire, physical constraints such as steep slopes and the narrow coastal shelf mean that the road connection stops some way short of the village.

[72] The Isle of Skye has been connected to mainland Scotland by the Skye Bridge since 1995. Causeways link the Hebridean islands of Eriskay, South Uist and North Uist, while several of the Orkney Isles are connected by roads built upon the Churchill Barriers, constructed during the Second World War in order to prevent German U-boat attacks on the UK Navy while stationed in Scapa Flow.

[73] In 2015, 707 billion passenger kilometres were recorded for road travel, 78 billion for rail and 8 billion for air. Department for Transport, Table TSGB0101: Passenger Transport by Mode, available at www.gov.uk/government/uploads/system/uploads/attachment_data/file/586451/tsgb0101.ods, last accessed 14 October 2017.

process and the planning approval process can be protracted and contentious.[74] Maintenance of routes of national significance (known as trunk roads) is the responsibility of the national highways authorities, while non-trunk roads are the responsibility of the relevant local authority. Many local roads, in particular, are in increasingly poor condition due to successive years of restricted local authority funding allocations from central government.

Owners of motor vehicles pay a road tax for permission to use the public roads, although the money raised thereby is not hypothecated for road maintenance. The amount payable depends upon the carbon emissions associated with the vehicle. Tolls are at present uncommon, although on one stretch of the M6 motorway drivers have the choice of paying a toll to avoid one of the more congested parts of the motorway network. Tolls are levied on some major bridges and tunnels, while congestion charges are payable in some cities.[75]

Reducing traffic accidents and the consequences thereof has been a major policy driver for successive UK governments. Measures such as the compulsory wearing of seatbelts as well as public awareness campaigns against drink-driving and the tightening of car safety regulations[76] helped to halve the annual number of fatalities from road accidents between 2000 and 2013.[77]

Road is the predominant means of moving freight within the UK, with the road network accounting for 170 billion tonne kilometres of movement in 2016[78] – more than five times the figure for domestic water-borne freight[79] and more than ten times the figure for rail.[80]

The railways play a significant role in the UK infrastructure system. Rail (including underground and metropolitan train travel) is one of the principal

---

[74] See, for example, *Walton v The Scottish Ministers* [2012] UKSC 44.

[75] At time of writing, London and Durham.

[76] Many of these measures have their roots in European law.

[77] Department for Transport, *Number of Fatalities resulting from Road Traffic Accidents in Great Britain*, available at www.gov.uk/government/publications/annual-road-fatalities, last accessed 14 October 2017.

[78] Department for Transport, *Domestic Road Freight Statistics 2016*, available at www.gov.uk/government/uploads/system/uploads/attachment_data/file/627597/domestic-road-freight-statistics-2016.pdf, last accessed 14 October 2017.

[79] In 2015, this amounted to 31.4 billion tonne kilometres, mainly comprised of port-to-port shipments. Department for Transport, *Domestic Water Borne Freight Statistics (Revised) 2015*, available at www.gov.uk/government/uploads/system/uploads/attachment_data/file/620296/dwf-2015-revised.pdf, last accessed 14 October 2017.

[80] In 2016 this amounted to 18 billion tonne kilometres. Department for Transport, Rail Trends Factsheet 2016, available at www.gov.uk/government/uploads/system/uploads/attachment_data/file/590561/rail-trends-factsheet-2016-revised.pdf, last accessed 14 October 2017. The decline in the movement of coal has significantly reduced the amount of freight transported by rail.

means of commuting to work in many of the UK's main conurbations and rail transport is one of the most commonly used means of inter-city travel. Service cuts made in the 1960s following the notorious Beeching Report[81] mean that rail plays a significantly more limited role in connecting rural communities than was once the case, although some lines remain and a modest number of previously decommissioned stations and lines have been brought back into service, generally to support commuter demand. The UK rail service was developed initially through private sector investment in the Victorian era. The rail service was nationalized in 1948, but privatized in stages throughout the 1990s as part of the wave of public sector flotations on the stock market which characterized the Thatcher government. The particular denationalization model used involved the separation of infrastructure ownership and service provision and the granting of passenger franchises. Not all of these ventures have been successful and the extent to which the government funds these franchises is a matter of some public debate and concern, as is the extent to which these services have ended up in foreign ownership.

Air travel is the most prevalent means of international travel to and from the UK, with 238 million passenger movements in 2015,[82] compared to 21 million for international ferries[83] and 10.5 million for the Channel Tunnel.[84] As has been noted above, air travel is less commonly utilized than either road or rail. Air travel does, however, help to connect island communities, with frequent scheduled services from the mainland to the Northern Isles and some of the Hebridean Isles.

The UK, as an island nation, has a long tradition of sea travel and shipping. The coastline is studded with ports of varying types and sizes.[85] Liquid bulk and dry bulk make up the majority of imports, although the volume of containerized freight is increasing.[86] Several UK ports are large enough to be able to

---

[81]   British Railways Board, *The Reshaping of British Railways*, London, HMSO, 1963.

[82]   Department for Transport, *Transport Statistics in Great Britain 2015*, available at www.gov.uk/government/uploads/system/uploads/attachment_data/file/489894/tsgb -2015.pdf, last accessed 14 October 2017.

[83]   Department for Transport, *Provisional Sea Transport Passenger Statistics 2015*, available at www.gov.uk/government/uploads/system/uploads/attachment_data/file/ 502712/prov-sea-passenger-statistics-2015.pdf, last accessed 14 October 2017.

[84]   Eurotunnel, *Traffic Figures*, available at www.eurotunnelgroup.com/uk/ eurotunnel-group/operations/traffic-figures/, last accessed 14 October 2017.

[85]   For a map, see Department for Transport, *UK Port Freight Statistics Revised (2016)*, available at www.gov.uk/government/uploads/system/uploads/attachment _data/file/646188/port-freight-statistics-2016-revised.pdf, p.17, last accessed 14 October 2017.

[86]   Ibid, p8.

handle containerized freight, with Felixstowe and Southampton between them accounting for more than half of the tonnage of such freight.[87] Some ports in the North of Scotland have focused on servicing the oil industry. Liquefied natural gas terminals and regasification facilities are to be found at South Hook and Dragon (both Milford Haven) and the Isle of Grain, Kent, with massive investment having been made in recent years in the Milford Haven facilities.

Ferries play a significant role in linking Northern Ireland to the rest of the UK and in connecting island communities to more populous islands and/ or to the mainland. Outside of these contexts, marine passenger transport is uncommon.

5.5.2   Energy Infrastructure

The UK has a highly developed downstream gas and electricity network allowing for the transmission and distribution of electricity and gas.[88] Virtually 100 per cent of UK households have access to mains electricity; however, the proportion of households with access to mains gas is significantly smaller and in some rural areas the majority of households are off the gas grid.[89] Upstream oil and gas infrastructure is in private ownership and is subject to a system of negotiated access.[90]

5.5.3   Access to Export Markets

At the time of writing, the UK is a member of the EU, having acceded to the European Economic Community on 1 January 1973 and having voted to remain a member in 1975. However, in a referendum held on 23 June 2016, the public voted to leave the EU. The process of leaving the EU was initiated when Article 50 of the Treaty of Lisbon was triggered on 29 March 2017 following the passage by Parliament of the Withdrawal from European Union (Article 50) Act 2017.[91]

---

[87]   Ibid, p16.

[88]   Ownership and operation arrangements are complex; see G Gordon, A McHarg and J Paterson, 'Energy Law in the UK', in M Roggenkamp et al, *Energy Law in Europe*, Oxford, 2016, para 14.161-165.

[89]   For a clickable map showing percentages of off-grid households by region, see National Grid, *Non-Gas Map*, available at www.nongasmap.org.uk/, last accessed 14 October 2017.

[90]   Op cit n 88, para 14.68–14.76.

[91]   The need for primary legislation in order to trigger the Brexit process was established in *R (on the application of Miller and another) v Secretary of State for Exiting the European Union* [2017] UKSC, 5, [2016] EWHC 2768.

Negotiations between the UK and the EU on the terms of the UK's departure have so far been progressing slowly and it remains very unclear what form that departure will take, or indeed even whether Brexit will happen. Withdrawal will take effect either when a withdrawal agreement enters into force or two years after notifying the European Council of the intention to withdraw. If there is no withdrawal agreement after two years and a veto on an extension period, or if the UK does not like the agreement, it can leave the EU without a withdrawal agreement.[92]

Given the multiple layers of uncertainty, it would perhaps be imprudent to say any more than that the UK remains a member of the EU in the meantime, and as a result has the benefit of free market access within the EU and the EU's negotiated trading relations with a variety of non-member states, while also being subject to European law and regulations.

The UK is not part of the Schengen accord for unrestricted movement of people across borders. The UK is not part of the Eurozone and has retained the pound sterling as its own national currency.

The UK is and has been a member of the World Trade Organization and its precursor, the General Agreement on Tariffs and Trade, since its foundation in 1947. Hopes have been expressed by some pro-Brexit campaigners that the UK's departure from the EU will permit it to enter into advantageous bilateral trade deals with a range of countries, including the US and members of the Commonwealth. At time of writing, however, it is not clear how quickly such deals are likely to be concluded or what their terms would be.

5.6   WELFARE SYSTEM

The UK has a 'mixed economy' welfare system, in which the state, the voluntary sector, the family and the market all play important roles; but the role of the state has increased since the 1870s.[93] In 2014–15 the UK government spent £258 billion on the welfare system, which was 35 per cent of all government spending. The discussion of the welfare system in this section is based on government expenditure on the welfare system and is divided by category and function.

---

[92]   House of Commons, *Brexit: How Does the Article 50 Process Work?*, available at http://researchbriefings.parliament.uk/ResearchBriefing/Summary/CBP-7551, last accessed 17 October 2017.

[93]   Bernard Harris, 'Health and Welfare' in Roderick Floud, Jane Humphries, and Paul Johnson, *The Cambridge Economic History of Modern Britain* (CUP 2014) 137.

5.6.1 Welfare System by Category

The welfare system in the UK is divided into six categories: (1) pensions; (2) family benefits, income support and tax credits; (3) incapacity, disability and injury benefits; (4) personal social services and other benefits; (5) housing benefits; and (6) unemployment benefits.

Pensions were the largest component of government spending, with a total share of £108 billion, in 2014–15. Spending on pensions has increased over the last five years, from 37.9 per cent in 2010–11 to 41.9 per cent in 2014–15. This increase reflects the country's growing and ageing population. In fact, the remaining life expectancy for an individual aged 65 in 2016 is 21 years for a man and 24 for a woman.

Since 2011, following the enactment of the Employment Equality (Repeal of Retirement Age Provisions) Regulation, the default retirement age was abolished, meaning that employers can no longer force employees to retire.[94] Before this law was introduced, 65 was the mandatory retirement age. Of course, in some cases following the implementation of this law, employers have had good reasons to force retirement, such as an individual working in a job that requires a certain physical condition. These employees can still claim their pension after reaching the state pension age or the age agreed by the provider in the case of a workplace pension.[95] Since 2010, the state pension age for women has been increased to reach parity with the male state pension age. In April 2017 the women's state pension age was 63 and nine months; it has been equalized with the men's state age of 65 since November 2018.[96]

In 2014–15, following pensions as the highest government expenditure, the government spent £44 billion on family, income support and tax credits. These include child benefits and support for people on low incomes. Around 3 million people were in in-work poverty in 2013. This means that their household income (adjusted for household size and composition) was below the poverty threshold. The 10 per cent of households with the lowest disposable income spent an average of £196 a week in 2013. Of this, half was spent on food and non-alcoholic drinks, transport, housing and household fuel and power. According to the UK government, income support is allocated to eli-

---

[94] The Employment Equality (Repeal of Retirement Age Provisions) Regulations 2011, SI 2011/1069, available at www.legislation.gov.uk/uksi/2011/1069/contents/made, last accessed 13 October 2107.

[95] GOV, *Working After State Pension Age*, available at www.gov.uk/working-retirement-pension-age, last accessed 13 October 2017.

[96] Pension Policy Institute, *Pension Policy Facts 2017*, available at www.pensionspolicyinstitute.org.uk/pension-facts/pension-facts-tables, last accessed 14 October 2017.

gible individuals – for example, pregnant women, carers, single parents with a child under five, those who are unable to work because of sickness or disability, persons between 16 and the pension credit qualifying age, and persons with no or low income and no more than £16 000 in savings.

The third highest share of government spending was on disability and injury benefits. In 2014–15 the government spent £41 billion on people who are ill or disabled. Disabled people are more likely to live in disadvantaged places and work in routine occupations. In the 2011 census, 18 per cent of people (10 million) reported some kind of disability. There is a wide range of disability-related financial support, including benefits, tax credits, payments, grants and concessions. In addition, the disabled get compensation if their injuries or disabilities were caused by serving in the armed forces.

Personal social services received 11.4 per cent of government spending in 2014–15, a slight decrease from 2010–11 (12.1 per cent). Personal social services cover a range of services provided by local authorities for many vulnerable groups, including the mentally and physically disabled, older people and neglected children. The expenditure for adult social services was about £17.04 billion. Spending on adult social care is mainly divided into three main areas: long-term care (£13 billion), short-term care (£587 million) and other social care (£3.43 billion). The combination of long and short-term support – a spending of around £10 billion – was spent on people aged 65 and over.[97] As for elderly care, in 2011 there were 9.2 million people aged 65 and over, making up 16 per cent of the UK's population.

The next category is housing benefits, on which the government spent £27 billion in 2014–15 – about 10 per cent of its total expenditure. People with low income can use housing benefits to pay part or all of their rent. The amount of financial aid depends on whether the applicant rents a private, council or social house. In each situation the amount depends on the eligible rent, spare rooms, household income and specific circumstances, such as the occupants' ages or disabilities.

In 2014–15, only 1 per cent, or £3 billion, of spending was used for unemployment benefits. The process of helping the unemployed comes in the form of the Jobseeker's Allowance (JSA). In January 2016, 760 200 people claimed these benefits – a decrease of 11.2 per cent compared with one year earlier. This benefit is for eligible people over 18 and under the state pension age who are available for work and actively seeking work, and who work on average less than 16 hours per week. Individuals can receive a maximum amount, but

---

[97] NHS Digital, *Personal Social Services: Expenditure and Unit Costs England 2015–16,* available at http://digital.nhs.uk/catalogue/PUB22240, last accessed 15 October 2017.

an individual's entitlement depends on age, income and savings. The JSA weekly amount for those aged 18–24 is £57.90; for those aged 25 or over it is £73.10 and for couples it is £114.85.

5.6.2   Welfare System by Function

The welfare system by function can be divided into: (1) social protection; (2) health; (3) education; (4) general public services; (5) defence; (6) public order and safety; and (7) other. From 1994 to the present, the three highest spending categories have been social protection, health and education. Social protection has remained one of the largest spending areas for the government and includes personal social services, pensions and benefits, equalling 38.1 per cent of total government spending.

Between 1994–95 and 2014–15, health spending steadily increased from 13.9 per cent to 19.8 per cent of all government spending. The UK is ranked sixth out of the G7 countries for healthcare expenditure as a proportion of GDP. In 2015 total healthcare expenditure was around £185 billion, of which 79 per cent (£147.1 million) was spent by the government. The total amount accounts for 9.9 per cent of the country's GDP. Education was the third largest area of government spending; proportionately, this has fallen slightly compared with 20 years ago. In 2014–15, education spending was 12.5 per cent of total government spending, while in 1994–95 it amounted to 12.7 per cent of total government spending. Spending on education is divided into four areas: education for children under five; primary education; secondary education; and tertiary education. Among these areas, secondary education has had the highest expenditure since 2010; in 2015 spending on this area reached £36.8 billion, or 44 per cent of the total budget. The government spent the least on education for children aged under five, spending only 6 per cent of the total budget.

5.7   CULTURAL PRACTICES

Hofstede (2010) has identified six dimensions for characterizing national cultures. The UK can be described as follows, where the index values (in brackets) range from zero to 100:

- Power distance (35): Low power distance. Society believes that inequalities should be minimized.
- Individualism (89): Highly individualist and private people. 'Self' is important, as is personal fulfilment, and people think about themselves and their contribution to society.

- Masculinity (66): A masculine society. Highly success oriented. People live in order to work and have clear performance ambitions. A competitive society.
- Uncertainty avoidance (35): Relatively comfortable in ambiguous situations and not threated by unknown situations.
- Long-term orientation (51): With an intermediate score of 51 in this dimension, a dominant preference in British culture cannot be determined. British society has links with the past and is dealing with the challenges of the present and future.
- Indulgence (69): A relatively high score in indulgence means people are willing to realize their impulses and desires with regard to enjoying life and having fun. Tendency towards optimism.

The Freedom of Information Act 2000 provides public access to information in two ways: first, public authorities are obliged to publish information and, second, members of the public are entitled to request information. To achieve a more open, transparent and accountable government, the UK prime minister in 2010–11 wrote two letters to all government departments and ordered them to publish their data on procurement, resources and finance in an open, reusable and regular format. In turn, the data.gov.uk website was launched in 2010, one of the largest open data resources in the world that brings together all government data and provides easy access for all people to track public spending, government expenditure and functions. In addition, through the contract finder option, the government and its bodies release details about contract and procurement tenders over £10 000.[98]

In 2016 the UK scored 81 on the corruption perception index (Transparency International). The index reflects the perceived level of public corruption, with a range from zero (extremely corrupt) to 100 (extremely non-corrupt). The UK scores tenth among the countries included in the index. High-ranking countries tend to have more press freedom, access to information about public expenditure, stronger standards for public officials and independent judicial systems. Although corruption in the UK is not endemic, some sectors are in danger, such as prisons, political parties, Parliament and sport. The UK Border Agency, police and prison service have been targeted by organized criminals. Social housing is exploited by organized criminals to facilitate drug trafficking and prostitution, or to house illegal immigrants who are involved in such activ-

---

[98] Cabinet Office, *2010 to 2015 Government Policy: Government Transparency and Accountability,* available at www.gov.uk/government/publications/2010-to-2015 -government-policy-government-transparency-and-accountability/2010-to-2015 -government-policy-government-transparency-and-accountability, last accessed 15 October 2017.

ities. In each of these areas the corruption of key officials, often in the form of bribery, is a critical factor in allowing the wrongdoing to occur.[99]

5.8   POLITICAL AND SOCIAL SUPPORT

Since 2015, 12 main parties have been represented in the House of Commons: the Conservative Party, the Co-operative Party, the Democratic Unionist Party, the Green Party, the Labour Party, the Liberal Democrats, Plaid Cymru, the Scottish National Party, Sinn Fein, the Social Democratic and Labour Party, the UK Independence Party and the Ulster Unionist Party. The system of political parties in the UK has existed since the eighteenth century. However, since the Second World War, all governments have been formed by only two parties: the Conservative Party and the Labour Party. The effectiveness of the party system, a significant factor in the Constitution, lies in the association between the government and opposition parties. The opposition party's goal is to participate in creating policy and legislation, opposing the government's proposals if it disagrees with them and promoting its own policies in order to win the next general election.[100]

The most successful party in the UK has been the Conservative Party, which has had two long terms in office, from 1951–64 and from 1979–97. Between 1997 and 2010, the Labour Party formed the government. However, in 2010 and 2017, the general elections produced a 'hung Parliament'. In 2010 the Conservative Party returned to power, but established a coalition government with the Liberal Democrats because it did not manage to win a majority. Likewise, in 2017 the Conservative Party formed a minority government with the Democratic Unionist Party because it did not gain an overall majority.[101] There are two variants of Conservatism, which are defined as 'One Nation' and 'Thatcherism'. The ideology behind 'One Nation' is to create social unity. 'Thatcherism' – the ideology inspired by former Conservative leader Margaret Thatcher – has three key strands: the free market, a small state and individual liberty and responsibility. Until 2010, the Labour Party was in office for 13

---

[99]   Transparency International UK, *Corruption in the UK: Overview and Policy Recommendations*, available at www.transparency.org.uk/publications/corruption -in-the-uk-overview-policy-recommendations/#.WeUGXK3MzYI, last accessed 16 October 2017.

[100]   Parliament, *The Party System*, available at www.parliament.uk/about/mps-and -lords/members/partysystem/, last accessed 15 October 2107.

[101]   House of Commons, *Hung Parliaments*, available at http://researchbriefings .parliament.uk/ResearchBriefing/Summary/SN04951, and House of Commons, *The 2010 Coalition Government at Westminster*, available at http://researchbriefings .parliament.uk/ResearchBriefing/Summary/SN06404, last accessed 16 October 2017.

years. The Labour Party's history dates back to the nineteenth century, when it attempted to represent the enfranchised working class. The Labour Party's approach is socialist, concentrating on equality.[102]

Home rule and devolution have been intermittently on the Scottish political agenda for over 100 years. In 1934 the Scottish National Party was founded and in 1974 it became the second largest party in Scotland in terms of vote.[103] Following the referendum on Scottish devolution in 1997, the Scotland Bill provided by the Labour government received Royal Assent in 1998 and become the Scotland Act, which established the devolved Scottish Parliament.[104] The Parliament compromised 73 constituency members who represent constituencies with the same boundaries as Scottish House of Commons seats (with the exception of Orkney and Shetland, which are individual seats).[105]

Alongside the Scottish attempt to achieve devolution, there have been efforts by some other UK islands to do the same. In particular, this has been the case with those endowed with abundant natural resources. For example, the Shetland movement emerged in 1978; following its establishment, it sought to achieve greater autonomy for Shetland and establish an assembly with legislative powers and control over direct taxation.[106] In 2015 a multi-party campaign group called Wir Shetland, which means 'our Shetland', was launched with the aim of gaining self-governing power. One of the most important aims and motivations behind such attempts is to obtain a fairer share of the oil revenues which have been extracted from their lands.[107]

Although the two main UK political parties have not explicitly discussed their differences in terms of the country's oil and gas industry, it can be understood from the Labour Party manifestos that it believes that the current pattern is not successful, fair or democratic, and that it needs to be refined. As such, it recommends government intervention and, in some cases, argues that state

---

[102] Alistair Clark, *Parties in the UK* (Palgrave 2012) 41–65.

[103] House of Commons Library, *Scotland and Devolution*, available at http://researchbriefings.parliament.uk/ResearchBriefing/Summary/RP97-92, last accessed 12 November 2017.

[104] Ibid.

[105] House of Commons Library, *Devolution in Scotland*, available at researchbriefings.files.parliament.uk/documents/SN03000/SN03000.pdf, last accessed 12 November 2017.

[106] Martin Dowle, 'The Birth and Development of the Shetland Movement: 1977–1980' (Scottish Government Year-books, 1981), available at www.era.lib.ed.ac.uk/handle/1842/9093, last accessed 12 November 2017.

[107] Wir Shetland, *Constitution*, available at https://wirshetland.org/wp/index.php/draft-constitution/, last accessed 12 November 2017.

ownership should be considered as a serious potential alternative.[108] Another main conflict between the two parties centres on shale gas. While the Labour Party has a plan to ban fracking,[109] the Conservative Party has declared that its priority is to expand shale gas.[110] Despite these sharp differences, it seems that the two main parties – at least in their general view – are aligned on increasing production from the North Sea (due to its positive impact on the UK economy and jobs), improving environmental awareness and accelerating the transition to cleaner energy.[111]

5.9   ENVIRONMENTAL STANDARDS

The Department for Environment, Food and Rural Affairs is the body responsible for the UK's natural environment. The department works with 33 agencies and public bodies, which are split into six main categories: the non-ministerial department; the executive agency; the executive non-departmental public body; the advisory non-departmental public body; the tribunal non-departmental public body; and other. The UK's offshore petroleum industry involves high-risk activities for the natural environment. The Department for Business, Energy and Industrial Strategy (BEIS) regulates all emissions and discharges from the UKCS.

Environmental considerations are divided into air pollution, climate change, flooding, land pollution, marine environment, noise and nuisance, waste water pollution, wildlife and countryside. There have been several significant Acts related to the environment, some of which are listed below in Table 5.2.

Poor air quality is known to be the greatest environmental hazard for the UK's public health. The proportion of adult mortality attributable to particulate air pollution was about 4.7 per cent in England in 2015.[112] The UK air

---

[108] The Labour Party, *Alternative Models of Ownership*, available at http://labour .org.uk/wp-content/uploads/2017/10/Alternative-Models-of-Ownership.pdf,   last accessed 19 October 2017.

[109] House of Commons, *Shale Gas and Fracking, 2017*, available at http:// researchbriefings.parliament.uk/ResearchBriefing/Summary/SN06073,   accessed   17 October 2017.

[110] Conservative Party, *Manifesto 2017, Forward Together: Our Plan for a Stronger Britain and a Prosperous Future*, available at https://s3.eu-west-2.amazonaws.com/ manifesto2017/Manifesto2017.pdf, last accessed 17 October 2017.

[111] The Labour Party, *Richer Britain, Richer Lives, Labour Industrial Strategy*, available   at   http://labour.org.uk/wp-content/uploads/2017/10/Richer-Britain-Richer -Lives-Labours-Industrial-Strategy.pdf, last accessed 17 October 2017.

[112] Public Health England, *Public Health Profiles*, available at https://fingertips.phe .org.uk/search/pollution#page/0/gid/1/pat/126/ati/102/are/E10000003, last accessed 19 October 2017.

Table 5.2   Environmental legislation

Air Pollution Acts

The Road Traffic Regulation Act 1984

The Clean Air Act 1993

The Environment Act 1995

The Finance Act 2000

The Air Quality (England) Regulations 2000

The National Emission Ceilings Regulations 2002

The Large Combustion Plants (National Emission Reduction Plan) Regulations 2007

The Environmental Permitting (England and Wales) Regulations 2010

The Air Quality Standards Regulations 2010

The Air Quality Standards (Wales) Regulations 2010

Land Pollution

Part 2A of the Environmental Protection Act 1990

The Contaminated Land (England) Regulations 2006

The Contaminated Land (Wales) Regulations 2006

DEFRA: Contaminated Land Statutory Guidance (April 2012)

The Environmental Damage (Prevention and Remediation) (England) Regulations 2015

The Environmental Damage (Prevention and Remediation) (Wales) Regulations 2009

Marine Environment Act

The Marine and Coastal Access Act 2009

The Environmental Damage (Prevention and Remediation) (England) Regulations 2015, SI 2015/810

The Environmental Damage (Prevention and Remediation) (Wales) Regulations 2009, SI 2009/995

The Environmental Damage (Prevention and Remediation) (Wales) (Amendment) Regulations 2015, SI 2015/1394

The Marine Strategy Regulations 2010, SI 2010/1627

Water Pollution Act

England and Wales

The Water Resources Act 1991

The Water Industry Act 1991

The Environment Act 1995

The Water Act 2003

The Water Environment (Water Framework Directive) (England and Wales) Regulations 2003

The Marine and Coastal Access Act 2009

The Environmental Permitting (England and Wales) Regulations 2010

The Water Act 2014

Scotland

The Water Environment and Water Services (Scotland) Act 2003

The Water Environment (Controlled Activities) (Scotland) Regulations 2005

The Marine (Scotland) Act 2010

The Water Environment (Controlled Activities) (Scotland) Regulations 2011

Climate Change Act
The Finance Act 2000 (as amended)
The Climate Change Levy (General) Regulations 2001
The Climate Change and Sustainable Energy Act 2006
The Finance Act 2006 (Climate Change Levy: Amendments and Transitional Savings in Consequence of Abolition of Half-rate Supplies) (Appointed Day) Order 2007
The Climate Change Act 2008
The Climate Change (Scotland) Act 2009
The CRC Energy Efficiency Scheme Order 2010
The Energy Act 2011
The Greenhouse Gas Emissions Trading Scheme Regulations 2012
The Energy Act 2013

Oil and Gas: Offshore Environmental Legislation
The Environmental Assessment of Plans and Programmes Regulations 2004
The Offshore Petroleum Production and Pipelines (Assessment of Environmental Effects) Regulations 1999 (as amended)
The Offshore Petroleum Activities (Conservation of Habitats) Regulations 2001 (as amended)
The Offshore Marine Conservation (Natural Habitats, &c.) Regulations 2007 (as amended)
The Offshore Chemicals Regulations 2002 (as amended)
The Offshore Petroleum Activities (Oil Pollution Prevention and Control) Regulations 2005 (as amended)
The Offshore Combustion Installations (Pollution Prevention and Control) Regulations 2013
The Greenhouse Gases Emissions Trading Scheme (ETS)
The Energy Savings Opportunity Scheme 2014
The Food and Environment Protection Act 1985, Part II Deposits in the Sea
The Energy Act 2008, Part 4A Consent to Locate
The Energy Act 2008 (Consequential Modifications) (Offshore Environmental Protection) Order 2010
The Pollution Prevention and Control (Fees) (Miscellaneous Amendments and Other Provisions) Regulations 2015
The Pollution Prevention and Control (Fees) (Miscellaneous Amendments) Regulations 2016
The Pollution Prevention and Control (Fees) (Miscellaneous Amendments) (No. 2) Regulations 2016

quality framework is derived from a mixture of domestic, EU and international legislation and consists of three main strands: legislation regulating total emissions of air pollutants, where the UK is bound by both EU law (the National Emission Ceilings Directive) and international law; legislation regulating concentrations of pollutants in the air, which is regulated by implementing the EU Air Quality Directive; and legislation regulating emissions from specific sources, using legislation that implements the Industrial Emissions Directive and the Clean Air Act. Although the UK voted to leave the EU in June 2016, the Repeal Bill will ensure that the whole body of EU environmental law will remain in UK law. The UK government has been relatively successful in reducing emissions of air pollutants over the past 70 years; for instance, it is estimated that between 2000 and 2016, emissions of nitrogen dioxide and particulate matter fell from 49 per cent to 26 per cent.[113] However, pollutant levels remain relatively high.

One environmental problem in the UK is acid rain, which is mainly caused by the burning of fossil fuels – oil, gas and coal – resulting in the acidification of upland waters and soil, creating a toxic environment for aquatic life. In the 1980s, controls on acidic emissions were introduced, which reduced the impact of acid deposition on soil, vegetation and surface waters. Further, the UK and other European countries have observed a dramatic reduction in the emission of gases into the atmosphere.[114]

Forests in the UK are a natural asset and an important part of the timber industry. In recent years, forests have come under risk from climate change, population growth, increasing pressure on land, diseases and pests. As a result of these risks, the UK has produced a plan to keep forests safe from these hazards. In addition to normal controls and restrictions, the measures adopted will ensure the health of trees and plants, and will manage risks posed by disease. In 2012, ash tree dieback was caused by the *Chalara fraxinea* fungus, discovered in the UK in Buckinghamshire. As 47 per cent of woodland is privately owned and not well managed, the Department for Environment, Food and Rural Affairs has implemented measures to regain management of woodland. In addition, the department attempts to stop deforestation and illegal logging around the world and encourages the use of sustainable palm oil. Three pieces of legislation are central to stopping illegal logging: the Forest Law Enforcement, the Governance and Trade Regulation and the EU Timber Regulation.[115]

Fisheries management in the UK has mainly been under the European Common Fisheries Policy (CFP). Once the UK leaves the EU, it means to withdraw from the CFP in an attempt to gain more control over fisheries and environmental policy. After Brexit, the UK has a plan to take back control of its fishing policy after 50 years of EU rule. The government will officially

---

[113] Department for Environment, Food and Rural Affairs, *Air Pollution in the UK 2016,* available at https://uk-air.defra.gov.uk/assets/documents/annualreport/air _pollution_uk_2016_issue_1.pdf, last accessed 19 October 2017.

[114] M Keman, RW Battarbee, CJ Curtis, DT Monteith and EM Shillalnd (eds), 'UK Acid Waters Monitoring Network 20 Years Interpretive Report' (The UK Upland Water Monitoring Network, July 2010), available at http://awmn.defra.gov.uk/ resources/interpreports/20yearInterpRpt.pdf, last accessed 19 October 2017.

[115] Department for Environment, Food and Rural Affairs, *2010 to 2015 Government Policy: Forests and Woodland,* available at www.gov.uk/government/publications/ 2010-to-2015-government-policy-forests-and-woodland/2010-to-2015-government -policy-forests-and-woodland, last accessed 19 October 2017.

begin withdrawal from the London Fisheries Convention that allowed foreign countries access to UK waters.[116]

5.10   COOPERATION BETWEEN INDUSTRY AND THE GOVERNMENT

Between 1981 and 2016, one million working days were lost in the UK, but this trend has been significantly decreasing over the last 30 years. In the 1970s and 1980s, the UK miners' strike (in 1972, by coal miners), the 'winter of discontent' (in 1979, by public sector workers) and the Battle of Orgreave (in 1984, by coal miners) caused a huge amount of working days to be lost. Prior to this, in the 1910s and 1920s, there were several strikes, such as the national coal strike (in 1912, by coal miners), the Battle of George Square (in 1919, by shipbuilding and engineering workers), Black Friday (in 1921, by coal miners) and the largest strike in UK history, the General Strike of 1926, which lasted nine days and was supported by more than 1.5 million workers.[117] However, in 2016, the main issue that resulted in working days lost related to the duration and patterns of working hours. The total number of working hours lost due to labour disputes was 322 000, mainly in the health and social work sector.[118]

Generally, conflict between an employee and his or her employer might centre on an employee's grievance or a concern of the employer. The grievance procedure, which is set out by employers, explains how a grievance decision can be appealed. Mediation, conciliation and arbitration are also available to solve conflicts before going to an employment tribunal. If the parties cannot resolve the problem, they can bring their dispute before an employment tribunal.[119] An employment tribunal (which, before 1998, was known as an industrial tribunal) is an independent judicial body that decides on disputes between employees and employers based on employment rights. The tribunal consists of three individuals: one representative of the employee, one representative of the employer and one qualified judge. The tribunals are considered a distinctive feature of UK administrative law, which aims to provide an equal,

---

[116] Department for Environment, Food and Rural Affairs, *The Government has announced it will withdraw from the London Fisheries Convention* (n 25).

[117] ONS, *The History of Strikes in the UK, September 2015*, available at http://visual .ons.gov.uk/the-history-of-strikes-in-britain/, last accessed 13 October 2017.

[118] ONS, *Labour Disputes in the UK: Analysis of UK Labour Disputes in 2016, Including Working Days Lost, Stoppages, and Workers Involved*, available at www.ons.gov.uk/employmentandlabourmarket/peopleinwork/workplacedisp utesandworkingconditions/articles/labourdisputes/latest, last accessed 13 October 2017.

[119] GOV, *Make a Claim to an Employment Tribunal*, available at www.gov.uk/ employment-tribunals/make-a-claim, last accessed 13 October 2017.

quick and more flexible procedure for the parties.[120] Employment tribunals hear claims about age, sex and race discrimination, equal pay, unfair dismissal, redundancy and so on. However, most claims relate to unfair dismissal.[121]

Trade unions have played a key role in advancing employees' rights. If a dispute between an employer and trade union members remains unsolved, the trade union can call for industrial action only if the majority of its members vote in favour through a properly organized postal vote.[122] The Employment Relations Act 1999 allows unions to apply for statutory recognition. Based on this Act, an employer might be compelled to recognize a union for collective bargaining on hours, holiday and pay. The law halted the efforts of employers to fail to recognize unions.[123] In recent decades, the number of workers who are union members has decreased. In 2016 around 6.2 million employees –23.5 per cent of the total workforce – were members of trade unions, which is considerably lower than the 13 million employees in 1979 who were members of trade unions. The proportions of female and UK-born employees who were in trade unions were higher than those of male and non-UK-born employees.[124]

In the 1970s, industrial strategy in Britain became synonymous with the failure of nationalized industries. The 1970s industrial strategy consisted of a number of policy mistakes and was characterized as a poor economy-wide decision, involving costly investment subsidies, misdirected research and development spending, protectionism, weak competition policy and a tax system that discouraged enterprise.[125] A move away from an industrial strategy policy happened in the late 1970s and 1980s, when free-market thinking (as

---

[120] Department for Business, Innovation and Skills, *The Employment Tribunals Rules for Procedure 2013 (as Subsequently Amended up to 17th February 2105)*, available at www.gov.uk/government/uploads/system/uploads/attachment_data/file/ 429633/employment-tribunal-procedure-rules.pdf, last accessed 13 October 2017.

[121] BEIS, *Understanding the Behaviour and Decision Making of Employees In Conflicts and Disputes at Work'*, Employment Relations Research Series No 119,2011, available at www.gov.uk/government/uploads/system/uploads/attachment_data/ file/32212/11-918-understanding-behaviour-employees-conflicts-at-work.pdf, last accessed 13 October 2017.

[122] GOV, *Taking Part in Industrial Actions*, available at www.gov.uk/industrial -action-strikes/your-employment-rights-during-industrial-action, last accessed 13 October 2017.

[123] Chris F Wright, 'What Role for Trade Unions in Future Workplace Relations?' (ACAS, September 2011), available at www.acas.org.uk/media/pdf/g/m/What_role _for_trade_unions_in_future_workplace_relations.pdf, accessed 13 October 2017.

[124] BEIS, *Trade Union Membership 2016: Statistical Bulletin*, available at www .gov.uk/government/uploads/system/uploads/attachment_data/file/616966/trade-union -membership-statistical-bulletin-2016-rev.pdf, accessed 13 October 2017.

[125] Centre for Economic Performance, *UK Growth: A New Chapter. A Blue Print For Growth in 2017 and Beyond*, available at www.lse.ac.uk/researchAndExpertise/

advocated by neoclassical economic thought) and policies of privatization and deregulation were introduced. However, an industrial strategy does not mean that the government directs or determines how industries operate. Instead, the government attempts to promote and sustain higher levels of productivity and economic growth. More recently, in 2017 the UK government returned to the industrial strategy policy perspective and advanced 10 key industrial areas in which to drive strategy across the entire economy, as follows: science; research and innovation; skills; infrastructure; business growth and investment; procurement; trade and investment; affordable energy; and sectorial policies.[126]

The current relationship between the UK government and the oil and gas industry is best understood by recent developments since the Wood Review Report. Following publication of the Wood Review Report on 24 February 2014, the OGA was established as an independent body with enhanced powers. Initially, it was an executive agency of BEIS. In October 2016 the OGA became a government company, with the Secretary of State for BEIS as its sole shareholder. The OGA is the main body that cooperates between the government and the industry – in essence, a stakeholder manager and project management administrative unit. Its purpose is to maximize the economic recovery of UK oil and gas through promoting investment in the UKCS, regulating the licensing of exploration and development of UK offshore and onshore oil and gas, and encouraging greater collaboration on the UKCS.[127]

5.11   CONCLUSION

The discovery of oil and gas in the North Sea occurred when the UK already had one of the most advanced economies in the world. The UK has always had a strong education system, and over the years has developed a diversified and relatively well-educated society. Strong parliamentary democracy is also a feature. The UK completely relied on the import of crude oil while its consumption was increasing, so the discovery of oil and gas in the North Sea was undoubtedly a major event in the post-war economic history of the UK. It made a significant contribution to UK economy in the form of jobs and tax revenue, and played a key role in the energy security of the country for some time.

---

units/growthCommission/documents/pdf/2017LSEGCReport.pdf, last accessed 14 October 2017.

[126] BEIS, *Building our Industrial Strategy*, available at https://beisgovuk .citizenspace.com/strategy/industrial-strategy/supporting_documents/buildingour industrialstrategygreenpaper.pdf, last accessed 14 October 2017.

[127] Oil and Gas Authority, *Our History*, available at www.ogauthority.co.uk/about -us/what-we-do/our-history/, last accessed 12 November 2017.

In contrast to the UK overall, the Shetlands established a special fund, the Shetland Charitable Trust, to receive and disburse the oil industry's revenues to the local community. Since its establishment in the 1970s, the Shetland Charitable Trust has redistributed over £300 million through charitable activities such as improving care for the elderly and building homes, and through investing in renewable energies.[128] In addition, the trust invested in the global markets or subsidiary companies in order to make more money for future activities.[129] Around 90 per cent of the trust's investments were made in the global markets and were valued at £231 million in 2017.[130] That said, as mentioned earlier, the people of Shetland would like to see a great redistribution of them.

Overall, for the UK – a country which used to import almost all of its oil and gas consumption – to reach 137 million tonnes of oil production at its peak in 1999 was unusual. However, there is a question mark over whether extraction has occurred at a sustainable level; and decommissioning in the North Sea may be very expensive and a cost that is passed on to the taxpayer. The issue of sustainability centres on whether the pace of extraction was too fast and fields were depleted too quickly. There is clearly an intergenerational distribution issue here, in terms of the financial revenue generated and the environmental effects from the extraction and use of these resources. Further, there are questions as to why the UK did not try to invest revenue in a similar way to the Shetlands.

This can be answered to some degree by the fact that in the decades since the discovery, the importance of different UK industry sectors has changed significantly. There has been a rapid shift towards the service sector, while many other sectors are in decline, so that some industries which used to have a considerable role in the economy no longer exist or now make negligible contributions to the economy. The UK has to some degree avoided the 'resource curse' – although it was never overly reliant on its oil and gas revenues, unlike many other nations which have suffered from the 'resource curse'.

In line with other sectors, the UK oil and gas industry has also experienced dramatic changes since the 1970s, facing challenges from the financial crisis of 2007–09 and the sharp decline in oil price since mid-2014 to the challenges arising from the current low levels of investment and the maturity of the UKCS. Nevertheless, the UK continues to view the North Sea as having the potential to make a positive economic impact, and has developed a new policy for its continued exploitation (Vision 2030) to develop the UK oil and gas sector and

---

[128] Shetland Charitable Trust, *Who We Are*, available at www.shetlandcharitabletrust .co.uk/who-we-are, last accessed 12 November 2017.

[129] Ibid.

[130] Shetland Charitable Trust, *Our Investments*, available at www .shetlandcharitabletrust.co.uk/our-investments, last accessed 13 November 2017.

its supply chain, generating an additional £290 billion to the UK economy by 2030.[131] These ambitious plans, however, may need to be tempered as decommissioning costs could be significant (perhaps between £80–£100 billion), and the issue of Brexit may create too much uncertainty for investors.

---

[131] Oil and Gas UK, *A Global Energy Industry, Anchored in the UK, Powering the Nation and Exporting to the World: Oil and Gas UK Blueprint for Government 2017*, available at http://oilandgasuk.co.uk/wp-content/uploads/2017/05/Blueprint-for -Government-2017.pdf, last accessed 26 October 2017.

6. Hydrocarbon policies and legislation: United Kingdom

Greg Gordon

6.1 INTRODUCTION

This chapter will outline the development of British hydrocarbon law and policy, and offer some critical comment relative thereto.[1] Focusing primarily upon the offshore industry, it will chart the evolution of British oil and gas law and policy; discuss the manner in which the state organizes its governance of the upstream oil and gas sector; outline the extent of British governmental participation; note the absence from Britain of oil funds; explain the extent and basis of the direct state take from upstream activities; and consider the means by which the British government seeks to encourage investment on the UK Continental Shelf (UKCS). Several major themes will be developed throughout: first, the relative lack of strategic planning and state direction in the British system, at least historically; second, the long shadow cast by decisions made in the early days of the development of the UKCS; third, the relevance of competing political ideologies at key moments in the development of the regime; fourth, the tendency – at least at certain key points – to see oil and gas as just an industrial sector like any other, and not as one requiring special treatment; and fifth, the potential for different limbs of the State to pull in different directions, causing adverse consequences for the regulated industry, and arguably, for the state itself. It will be seen that (at least outside the field of tax) the cumulative effect of these factors had, until recently, been the creation of a system of non-interventionist governance where the State plays a less pronounced policy role than is the case in many other oil and gas provinces. However, in the Wood Review we have seen a recent acceptance of the fact that this approach

[1]   The author would like to thank the editors and reviewers, as well as Dr Emre Üşenmez and Profs John Chandler, Ernst Nordtveit and Hector MacQueen for their valuable comments on an earlier draft of this work. However, none of the above bear any responsibility for any errors and omissions that remain.

may not be sufficient to meet the challenges of a rapidly maturing province, and the implementation of significant regulatory reform.

6.2   HISTORICAL DEVELOPMENT OF BRITISH HYDROCARBON LAWS AND POLICIES

6.2.1   The Evolution of British Hydrocarbon Law

The first British[2] oil industry did not involve the search for hydrocarbons in their conventional form, but the use of a process of destructive distillation to produce oil, initially from coal and latterly from oil shale.[3] The story of the rise and fall of this industry is a fascinating one;[4] however, the volume of oil produced in this manner was never large and, after peaking in the mid-nineteenth century, the industry went into decline at the time of the First World War, when cheap crude began to be imported in significant quantities from Persia; it continued in a diminished form for some time before finally coming to an end in 1962.[5] For present purposes, we will confine ourselves to a consideration of the laws and policies that have attended the exploration and production of crude oil and natural gas. Generally, the focus of the work shall be offshore development, but where appropriate – for instance, in the following examination of the onshore origins of the offshore system of licensing – onshore developments shall be considered.[6]

---

[2]   'Great Britain' is made up of England, Wales and Scotland. The 'United Kingdom' is comprised of Great Britain and Northern Ireland. The devolved assembly of Northern Ireland has legislative competence over onshore oil and gas matters within that jurisdiction, and a separate licensing system exists there, albeit one that is little used: see G Gordon, A McHarg and J Paterson, 'Energy Law in the UK', in M Roggenkamp, C Redgwell, A Ronne and I Del Guayo, *Energy Law in Europe*, 3rd ed, 2016, at 14.11. When discussing onshore development, this chapter focuses on the law as applies in Great Britain and as a result the terms 'Britain' and 'Great Britain' will generally be used in preference to 'the United Kingdom'. However, when discussing the offshore area, the expression 'United Kingdom Continental Shelf' (UKCS) will be used, as the Northern Irish assembly has no legislative competence in the offshore area.

[3]   Similarly, gas was manufactured by destructive distillation from coal and used, by means of local distribution networks, for heating and light long before the large-scale production of natural gas.

[4]   For an account, see J MacKay, *Scotland's First Oil Boom; The Scottish Shale-oil Industry 1851–1914*, John Donald Press, 2012.

[5]   B Harvie, 'Historical Review Paper – The Shale Oil Industry in Scotland, 1858–1962, Part I: Geology and History', 27 (2010) *Oil Shale*, 354–358 at 357.

[6]   For an account of the current law relating to onshore unconventional developments, see the sources referred to at n116, below.

6.2.1.1   The Petroleum (Production) Act 1918

Winston Churchill's decision, taken when appointed First Lord of the Admiralty in 1911, to switch the British naval fleet from coal to oil[7] acted as a spur to secure supplies of crude oil.[8] This led to Britain first taking steps to secure overseas supply[9] and second, attempting to develop a domestic onshore extractive oil industry. During wartime, this attempt was governed by regulations promulgated in terms of the Defence of the Realm Acts,[10] and the Petroleum (Production) Act 1918 ('the 1918 Act') implemented Britain's first peacetime legislative framework for exploration and production of petroleum. A previous attempt had been made, but the issue had proven to be controversial. A previous Bill, which had sought to vest property rights in the State, was withdrawn after meeting with strident opposition from large-scale British landowners, who contended that petroleum present beneath land was owned by the owner of the overlying land, in like fashion to most other mineral rights.[11] They opposed the notion of the State making a claim to petroleum in strata without the payment of compensation.[12]

The 1918 Act did not directly address this controversy. Instead, the government deliberately side-stepped the issue of who owned oil and gas deposits,[13] leaving the question unresolved.[14] The Act's provisions were based not upon

---

[7]   By 1913, the process was well underway; W Brewer, 'Yesterday and Tomorrow in the Persian Gulf', 23 (1969) *Middle East Journal* 149 at 152.

[8]   D Yergin, *The Prize: The Epic Quest for Oil, Money and Power*, 1991, Free Press, pp11–12.

[9]   See (on the benefits of oil over coal naval power) W Churchill, HC Deb 26 March 1913 vol 50 col 1771 and (on the government's plans to secure overseas supplies) W Churchill, HC Deb 17 July 1913 vol 55 cols 1474–5. For a short account of British-Persian oil relations, see M Rubin, '*Stumbling Through the Open Door – The US in Persia and the Standard-Sinclair Oil Dispute*', 28 (1995) *Iranian Studies*, pp203–229 at pp203–206.

[10]   See C Cook (ed), *Defence of the Realm Manual*, 6th ed, 1918, HMSO pp43–44.

[11]   That is, the Petroleum Bill 1917–18. By means of a parliamentary answer given on 14 January 1918, the government confirmed that it was withdrawing the Bill: HC Deb 14 January 1918 vol 101 col 24.

[12]   HC Deb 17 July 1918 vol 108 cc1045–6, where Captain Wright, Member of Parliament for Leominster, warned the government that the question of 'confiscation' was 'highly contentious'.

[13]   The Minister introducing the Bill, stated: 'There was a previous Bill introduced last Session, but some of the proposals of that Bill were objected to and it was not thought right to attempt to press a controversial question as a War measure. This Bill avoids, I hope, all controversial questions.' HC Deb 01 August 1918 vol 109 col 620.

[14]   In a written answer to a question posed shortly after into the entry into force of the 1918 Act, the Attorney General would not commit himself to a position on who owned sub-surface oil. Describing it as 'a difficult question, upon which opinions may easily differ', he indicated that he personally inclined to 'the view that the surface

considerations of property law, but upon the precept of administrative control. It provided that only those holding a licence from His Majesty could 'search or bore for or get' petroleum, and that anyone undertaking such works without authority would forfeit any petroleum so obtained and in addition pay a penalty of three times its value.[15]

The 1918 Act did not lead to the development of a large-scale domestic British oil extraction industry. By the time of the introduction of the Petroleum (Production) Act 1934, the government would recognize that the British petroleum industry had been held back by the uncertainty surrounding the question of who owned oil in strata. In the Second Reading of the Petroleum (Production) Bill 1934, the President of the Board of Trade described the proprietary issues surrounding the ownership of petroleum as 'a complicated and increasing obstacle' which it was 'essential' to resolve. He continued:

> we have come to the conclusion that the only way in which it is possible to simplify matters and to remove the obstacles to the operations of the oil properties which would give any chance of adding to the national wealth and national security by means of petroleum, is to vest those oil properties, if they exist at all, in the Crown.[16]

While the problems caused by the property issue were real, it is worth noting that other practical reasons are also likely to have contributed to the failure to develop a domestic British oil industry. The state-funded drillings undertaken immediately after the passage of the 1918 Act proved to be largely unsuccessful and provided little encouragement for other landowners to drill.[17] Moreover, Britain's success in securing a regular and dependable supply of petroleum at modest cost meant that there was little economic incentive or policy imperative to drive the development of a domestic industry.

---

owner is the owner of the oil beneath his land in the sense that he alone has the right on his land to search and bore for oil and becomes the owner of the oil if and when he gets it'. He indicated that any such right was, however, subject to the provisions of the Petroleum (Production) Act 1918: HC Deb 12 August 1919 vol 119 col 1128W, available online at http://hansard.millbanksystems.com/written_answers/1919/aug/12/british -oilfields-surface-owners#S5CV0119P0_19190812_CWA_104, last accessed 19 April 2019. With respect, this attempt to evade the question shows considerable conceptual confusion, as it manages to suggest both that the surface owner is (at least in a sense) the owner of the sub-surface oil and that the rule of capture applies.

[15] Petroleum (Production) Act 1918, s 1(1).
[16] HC Deb 19 June 1934 vol 291 col 216.
[17] HC Deb 22 March 1934 vol 287 cols 1381–2 per the President of the Board of Trade.

6.2.1.2 The Petroleum (Production) Act 1934

As has been noted elsewhere, '[t]he Petroleum (Production) Act 1934 fundamentally altered the nature of the British petroleum licence'.[18] Section 1(1) of the 1934 Act expressly vested in the Crown 'property in petroleum, existing in its natural condition in strata in Great Britain'. The Crown was also granted an exclusive right to search and bore for and get such petroleum.[19] As owner of the petroleum, the Crown exercised the right to extract a financial benefit in exchange for permitting the exploitation of its asset; initially this was done by means of royalty, although as we shall see, this has since been supplanted by tax.[20] The 1934 Act did not provide for penalties in the event of breach but, as it had effectively moved the British system away from the administrative exemption from punishment model to a property-based system, penalties were no longer necessary, at least in order to protect the state from unauthorized production.[21] This schema, as re-enacted in the Petroleum Act 1998, continues to form the basis of the current licensing regime.

The government asserted that the Crown's assumption of rights of ownership over oil in strata did not amount to expropriation; or at least, did not amount to the expropriation of a valuable asset. In the Second Reading of the Petroleum (Production) Bill 1934, the President of the Board of Trade stated: 'we hold the view that these rights, if they exist at all, are purely imaginary and have no practical value, and never have had any practical value.'[22] This position was endorsed by the opposition,[23] which voted with the government, but was met with some dismay by some of the governing party's own back-benchers, who spoke – unsuccessfully – in support of the rights of the landowner.[24] Consistent with the fact that the government did not accept that persons owning the land overlying deposits of oil had any meaningful commercial rights in the oil

---

[18] G Gordon, 'Petroleum Licensing', in G Gordon, J Paterson and E Üşenmez, *Oil and Gas Law: Current Practice and Emerging Trends*, 2nd Ed, DUP, 2011 ('Gordon, Paterson and Üşenmez') para 4.7.

[19] Petroleum (Production) Act 1934 s1(1).

[20] See n211 and accompanying text.

[21] As owner, if an unlicensed party sought to withdraw oil, the State would be entitled to rely upon the normal remedies available under property law.

[22] HC Deb 19 June 1934 vol 291 cc213–319, available at http://hansard .millbanksystems.com/commons/1934/jun/19/petroleum-production-bill-lords #S5CV0291P0_19340619_HOC_341, last accessed 19 April 2019, per The President of the Board of Trade (Mr Runcitnum) at 216.

[23] The opposition was delighted with the expropriation, but far from content with the government's proposal to open the industry up to private enterprise; the opposition's preference was for a nationalized oil industry. Ibid per Mr Hall at 221.

[24] Ibid, per the Marquess of Hartington, at 230. The Marquess's intervention discloses that he himself was a landowner who had oil beneath his land.

itself, the 1934 Act did not provide for compensation for the owners of land overlying petroleum deposits.[25] It did, however, provide, by the extension to petroleum operations of the provisions of the Mines (Working Facilities and Support) Act 1923, a system of compensation for loss of amenity associated with the workings undertaken to drill for and get the oil.[26] The idea that a right is only being taken away if it is of proven commercial value[27] is, from the standpoint of principle, deeply problematic; and given the scale of the world-wide petroleum industry and the importance that parties attach to securing a modest increase in the share attributable to their joint operating agreement (JOA) group in a unitized field,[28] the argument that rights in petroleum are of imaginary or nugatory value looks very peculiar to modern eyes. Nevertheless, the government was able to command a comfortable majority on the Bill.[29]

The clarification of the question of who owned oil in situ was not in itself sufficient to lead to the development of a large-scale onshore petroleum indus-try. The volume of production remained broadly static at pre-1934 levels until the end of 1960s.[30] It began to creep up throughout the first half of the 1970s, before spiking sharply upwards with the entry into production of the Wytch Farm field on the Dorset coast – a substantial field of comparable size to many offshore finds which has provided the overwhelming majority of Great Britain's onshore oil production.[31]

---

[25]   In this it may be contrasted with, for example, the nationalization of the coal industry, where compensation was payable following expropriation: Coal Industry Nationalisation Act 1946 s 10.

[26]   The successor statutory regime, contained in the Mines (Working Facilities and Support Act) 1966 s 8(2), formed the basis of the court's determination of the compen-sation payable when an oil company committed a subterranean trespass when drilling for oil beneath another's land: *Bocardo SA v Star Energy UK Onshore Ltd and another* [2010] UKSC 3. For a discussion of the broader issues arising in the case, see C Stacey, *Case Comment: Bocardo SA v Star Energy UK Onshore Ltd* 2 (2011) ASLR 124.

[27]   See, for example, Kingsley Griffith op cit n 23, 227. He recognized that prop-erty may exist as 'an abstraction', but that this should not be 'guarded and respected'; instead, it could (and on this occasion, should) be removed before 'vested interests' had grown up.

[28]   P Worthington, *Provision for expert determination in the unitization of strad-dling petroleum accumulations*, 9 (2016) JWELB, 254–268, 260.

[29]   In its second reading in the House of Commons, the Bill passed by 251 votes to 29.

[30]   For the decade prior to 1934, production was consistently between 115 and 170 thousand tonnes per annum. In the period from 1934 to 1969, it was between 79 and 235 thousand tonnes; see www.gov.uk/government/uploads/system/uploads/attachment _data/file/447442/Oil_Production___Trade_since_1890.xls, last accessed 18 April 2019.

[31]   See the data cited ibid, read together with the data contained at BEIS, Oil Production since 1975, available at www.gov.uk/government/uploads/system/uploads/ attachment_data/file/434575/Appendix9.xls, last accessed 18 April 2019.

6.2.1.3   The advent of offshore exploration and production

Exploration in the UK's offshore area was prompted by a combination of discoveries made by its near-neighbours and geopolitical threats to the security of its supplies of imported oil. The realization in the late 1950s that geological formations associated with the giant Groningen gas field in the Netherlands extended under the North Sea led to considerable exploration activity in the southern part of the North Sea. In due course, gas was discovered by the Sea Gem, a customized jack-up barge,[32] in what was to be known as the West Sole field in September 1965.[33] In 1969 the discovery of the giant Ekofisk field[34] upon the Norwegian Continental Shelf would serve to intensify the search for oil further north in the UKCS; and the State and oil companies' concerns about the high costs of exploring and producing in what was then a frontier province were diminished by the oil shocks caused by the major producing nations' use of the 'oil weapon'[35] in 1967 and in particular 1973.[36]

When the UK laid claim to the UKCS and sought to promulgate a legal framework authorizing the exploration for, and production of, offshore oil and gas, it did so by exporting into the seaward area the existing but largely untested landward licensing regime. Professor Alex Kemp's *Official History of North Sea Oil and Gas* makes it clear that little thought was given to any alternative method of developing an offshore licensing regime, and that little effort went into the process of trying to identify material differences between onshore and offshore development.[37]

Section 1(1) of the Continental Shelf Act 1964 vested in the Crown 'any rights exercisable by the United Kingdom outside territorial waters with respect to the sea-bed and subsoil and their natural resources', except coal,

---

[32]   Scant months later, the Sea Gem would be associated with tragedy when it collapsed, leading to the deaths of 13 persons onboard. See further J Paterson, 'Health and Safety at Work Offshore', in G Gordon, J Paterson and E Üşenmez, *Oil and Gas Law: Current Practice and Emerging Trends*, 2nd ed, DUP, 2011, paras 8.6–8.10.

[33]   J Bamberg, *British Petroleum and Global Oil 1950-1975: The Challenge of Nationalism*, CUP 2000, 201.

[34]   The massive Ekofisk field was Norway's first producing field and continues to be highly productive. See the information available at www.conocophillips.no/EN/our-norway-operations/greater-ekofisk-area/ekofisk/Pages/default.aspx, last accessed 18 April 2019.

[35]   A Alhajji, 'The Oil Weapon: Past, Present and Future', 17 (2005) *Oil & Gas Journal* 22.

[36]   T Priest, 'Shifting Sands: The 1973 Oil Shock and the Expansion of non-OPEC Supply', in E Bini, G Garavini and F Romero (eds) *Oil Shock: The 1973 Oil Crisis and Its Economic Legacy*, I B Tauris, 2016, 118.

[37]   A Kemp, *Official History of North Sea Oil: Vol I: The Growing Influence of the State*, Routledge, 2012, (Kemp, *Official History: Vol 1*) p9.

where the right was vested in the National Coal Board.[38] The effect of this section was to lay claim to the 'sovereign rights' provided by international law to natural resources located upon the UKCS.[39] Section 1(3) of the 1964 Act applied most of the key licensing provisions of the 1934 Act to the UKCS. However, s 1 of the 1934 Act (vesting of property in petroleum *in strata*) was not so applied, presumably as a full claim of ownership would be inconsistent with the claim to sovereign rights. Precisely how such a sovereign right differs from a right of full ownership remains a matter of some disagreement, more than 50 years on;[40] however, it is clear that something less than a full right of ownership of the seabed is involved.[41]

Perseverance with the model provided by the 1934 Act meant that the provisions of the petroleum legislation (and now applicable both onshore and offshore) continued to be skeletal in nature, providing broad discretionary powers to the Secretary of State and leaving the fine detail of the relationship between the State and the licensee for the Model Clauses[42] and the terms of the individual licence – an instrument which is contractual in form, but regulatory in nature.[43] As we shall see in section 6.4, state entities participated in these early licences alongside private enterprise, albeit in a somewhat ad hoc manner.

---

[38] Continental Shelf Act 1964, s 1(2).

[39] Contained, at the time, in the United Nations Convention on the Continental Shelf 1958, Art 2(1); now United Nations Convention on the Law of the Sea 1983, Art 77(1).

[40] G Gordon, 'Petroleum Licensing', in Gordon, Paterson and Üşenmez, op cit n 18, paras 4.8–4.9 and the sources cited therein; T Daintith, in T Daintith, G Willoughby and A Hill (eds), *United Kingdom Oil and Gas Law*, (1984-date, looseleaf, Sweet and Maxwell) ('Daintith, Willoughby and Hill'), paras 1–345 to 1–346.

[41] S Jayakumar, 'The Continental Shelf Regime under the UN Convention on the Law of the Sea: Reflections after Thirty Years', in M Nordquist and others (eds) *Regulation of Continental Shelf Development* (Martinus Nijhoff 2013), at 6.

[42] The Model Clauses – essentially standard terms and conditions – are set forth in delegated legislation and then incorporated into the licence. As I have noted elsewhere, the licence incorporates the set of Model Clauses current at the time of the licence's grant. This means that the production licences current on the UKCS are held on a variety of different terms. This, together with the fact that the State has no right to unilaterally amend the licence, has complicated the system and led to difficulties in implementing legal change. See G Gordon, 'Production Licensing on the UK Continental Shelf: Ministerial Powers and Controls', 4 (2015) *LSU Journal of Energy Law and Resources*, pp75–95 at pp77–78.

[43] Scanned copy licences are available to view at https://itportal.decc.gov.uk/web _files/recent_licences/oglicences.htm, last accessed 23 May 2017. Upon review, it can be seen that they have much of the form of a commercial contract: they are made 'between' the State and the licensee, and the fact that consideration is passing in respect of the grant is expressly narrated. However, they are regulatory in nature, imposing a broad range of obligations upon the licensee.

The early licensing and governance regime may, with the benefit of hindsight, be characterized as insufficiently sophisticated to properly protect the State's interest. Licences were granted with long terms, minimal relinquishment requirements and limited powers on the part of the State to direct or influence the pace of operations.[44] It was also possible for a licensee to manipulate its tax affairs so as to artificially create a situation where no tax was payable relative to its upstream operations.[45] Some of these problems would be addressed in the mid-1970s with the Petroleum and Submarine Pipelines Act 1975 (which provided for, among other things, more power for the State to influence the pace of exploration and production activities)[46] and the Oil Taxation Act 1975 (which established a bespoke system of petroleum taxation).[47] Other problems would continue to dog the industry well into the millennium.[48]

---

[44] See further G Gordon, 'Production Licensing on the UK Continental Shelf: Ministerial Powers and Controls', 4 (2015) *LSU Journal of Energy Law and Resources*, pp75–95.

[45] The Committee of Public Accounts produced an insightful and influential report in 1973 which noted this weakness in the UK's fiscal system: First Report from the Committee of Public Accounts, Session 1972–1973: North Sea Oil and Gas (HC 122). See also Kemp, *Official History: Vol 1*, pp249–251.

[46] Petroleum and Submarine Pipelines Act 1975 s 17. Not only were these conditions imposed prospectively; they were retrospectively incorporated into all existing offshore licences: ibid, s 18. The retroactive application of the provisions, in particular, proved controversial and led to a heated debate in Parliament. In seeking to defend their actions, the Labour government contended that the problem that the provisions sought to address had been caused by the previous Conservative government, which had failed to protect the nation's interests, and had instead 'given away' large tracts of the UKCS 'like Green Shield stamps': Hansard, HC Deb 30 April 1975, 5th series, vol 891, col 555 per Mr Canavan. For the Conservatives, Labour's reforms amounted to changing the rules of the game after it had commenced, without offering compensation to the companies affected by the change: ibid cols 503–504 per Mr Jenkin.

[47] Oil Taxation Act 1975. Part I thereof introduced Petroleum Revenue Tax; s 13 introduced Ring-fence Corporation Tax. For an account of these taxes, see E Üşenmez, 'The UKCS Fiscal Regime' in Gordon, Paterson and Üşenmez op cit n18, 137–160.

[48] The grant, in the first four licensing rounds, of licences with long terms and limited relinquishment provisions created difficulties that were not adequately dealt with until the introduction of the Fallow Fields initiative. See further G Gordon and J Paterson, "Mature Province Initiatives", in G Gordon, J Paterson and E Üşenmez, op cit n18, paras 5.7–5.11.

6.2.2   The Evolution of British Hydrocarbon Policy

6.2.2.1   Pragmatic discretion and reaction to events

Given how underdeveloped the upstream British oil industry was prior to the beginning of the offshore era, there is little purpose in discussing policy prior to the late 1960s. Indeed, if by 'policy' is meant a structured and logically coherent plan for the development of oil and gas operations within the province, it could be argued that it is only with the advent in 2015 of the Oil and Gas Authority (OGA) that it becomes meaningful to talk of oil and gas policy upon the UKCS. As we shall see further below, Britain has not shied away from setting high-level aims such as the desire to 'encourage rapid and thorough exploration and economical exploitation';[49] however, detailed strategy on how to achieve these goals has until very recently been lacking.[50] There are a number of reasons for this. First, and perhaps most obviously, it is naturally very difficult to establish a detailed strategy without having significant knowledge and experience of the matter in hand; states lack this expertise when embarking upon the pioneering phase of upstream oil and gas operations. But while this can potentially explain the lack of a coherent strategy at the outset, it does not account for the failure to develop a strategy as the government's knowledge base and expertise expanded. A deeper reason or reasons must therefore exist to explain the ongoing omission. Tradition[51] or regulatory culture may provide at least part of the answer. As Nelsen notes, British civil servants were much more inclined to adopt an arm's-length approach to the regulation of industrial sectors than were their Norwegian counterparts.[52] Moreover, British companies had been significant players in the international oil and gas industry for some time. One might have expected this to be a good thing, providing a well of experience from which Britain could draw; however, it also introduced an element of complication into the administrative process, with the government at least initially unwilling to take steps (eg, the introduction of a state national oil company) that could have an adverse effect upon the valuable interests of existing British-based international oil companies.[53]

---

[49]   FJ Erroll, Conservative Minister for Power, quoted in K Dam, *Oil Resources: Who Gets What How?*, University of Chicago Press, 1976, at 25.

[50]   See further the discussion on 'maximizing economic recovery', below.

[51]   "The main outlines of the British system for exploitation were undoubtedly determined as much by tradition as by logic": Dam, op cit n49, p23.

[52]   B Nelsen, 'Explaining Petroleum Policy in Britain and Norway 1962–90', 15 (1992) *Scandinavian Political Studies* 307–326, ('Nelsen, "Petroleum Policy in Britain and Norway"') p315.

[53]   'The British, with their extensive political and commercial involvement in the Middle East, did not want to set a precedent harmful to British interests by creating an onerous licensing system at home.' Nelsen, 'Petroleum Policy in Britain and Norway'

For instance, there were concerns that compelling a company such as BP to enter into a joint venture with a British state national oil company might make it practically difficult for BP to insist upon a sole concession in its negotiations with other foreign powers.[54] Last but by no means least, post-war Britain was in a parlous economic state when the offshore search for oil and gas commenced[55] – a fact which was not conducive to an extended period of reflection and policy development prior to the commencement of the search for hydrocarbons. The approach of government was to get the industry going and to pragmatically resolve problems as and when they were encountered. So the highly discretionary and underdeveloped onshore licensing system was exported offshore with minimal thought as to whether another approach might be preferable;[56] and when, in the parliamentary debates relative to the introduction of the Petroleum and Submarine Pipe-lines Act 1975 Act, the then-Labour Under-Minister for Energy chided his Conservative predecessor for failing to adopt measures to control depletion,[57] the response was that someone in his position 'would have had to be far-sighted indeed'[58] to imagine such things necessary. In other key areas – tax,[59] health and safety,[60] third-party access

---

p315. As we shall see below, in time and with changes to the political complexion of the government, policy on the desirability of a state national oil company would change and change again.

[54] Kemp, *Official History: Vol 1*, n37, above, p93.

[55] See, for example, Sir Alec Cairncross, 'Devaluing the Pound: The Lessons of 1967', *The Economist*, 14 November 1992, 25–28.

[56] See n37 and associated text, above.

[57] HC Deb 30 April 1975 vol 891 cc482-611 per Mr Smith at col 505. Depletion policy is concerned with the notion that as a barrel of oil is produced only once, it should be produced at the time of maximum advantage to the state and was a significant feature of Labour's 1975 reforms. See further Kemp, *Official History: Vol 1*, n.37, above, pp349–352.

[58] Ibid per Mr Jenkin at col 506.

[59] As already noted, the need for a specialist tax regime was not identified until 1972/1973: see n45, above.

[60] The early approach to health and safety regulation was little short of shambolic. Deep confusion and disagreement about which onshore provisions should be extended offshore eventually gave way to the realization that none of the measures under discussion were particularly apposite: Kemp, *Official History, Vol 1*, 20–21. This led to a 'quick fix', whereby health and safety was regulated by means of a single licence condition and the incorporation by reference of industry best-practice guidance; however, this had not been drafted with a view towards regulatory effect and the attempt was inept: Paterson, *Health and Safety at Work Offshore*, para 8.5–8.10. The system was amended in the light of the Sea Gem disaster, but it was not until after the Piper Alpha disaster that it emerged in its modern form: Paterson, *Health and Safety at Work Offshore*, paras 8.27-8.50.

to infrastructure[61] and decommissioning[62] – problems would be encountered which would expose the primitive nature of the initial British approach. In each of these areas we have seen significant improvement in the UK's laws and regulation. There are other areas where serious problems have not yet arisen where flaws are waiting to be exposed.[63] Historically, the UK's approach could also be characterized as passive or reactive in the sense that the government has generally considered its role to be to act as a permitting authority, reacting to and evaluating individual requests for acreage from the private sector rather than imposing its strategic vision upon the industry.[64] It has also been focused upon measures directed towards the individual licence or asset, not a region or province as a whole. However, with implementation of the Wood Review's proposals now underway, we would seem to be set to enter into a new era of more interventionist and holistic governance.

6.2.2.2 The relevance of political ideologies

Throughout the late 1960s and 1970s, policy changed markedly in accordance with the political ideology of the governing party. This was particularly true in the context of the approach to direct state participation,[65] but was also evident in other areas – for instance, the extent to which the State should retain a power to direct or control exploration and production activities (or should simply trust the industry to act in its own enlightened self-interest).[66] Particularly in the

---

[61]   The system of third-party access to infrastructure has developed in a piecemeal and unsatisfactory manner and has been materially held back by the relevant department's unwillingness to develop a detailed and coherent strategy and set of clear norms governing third-party access. Instead, the relevant department has clung to an approach that requires the exercise of considerable discretion, inhibiting certainty and impeding timeous development. See Gordon, McHarg and Paterson, 'Energy Law in the UK', para 14.72.

[62]   Prior to the entry into force of the Petroleum Act 1987, serious lacunae existed in the offshore decommissioning regime.

[63]   For example, the law of civil liability for pollution caused by a release of oil from an offshore platform. See, for example, G Gordon, 'Oil, Water and Law Don't Mix', Parts 1 and 2, 25 (2013) *Environmental Law and Management*, pp 3–11 and 121–132; M Faure and H Wang, 'Compensating victims of a European Deepwater Horizon accident: OPOL revisited', 62 (2015) *Marine Policy*, pp25–36.

[64]   See further G Gordon, 'Production Licensing on the UK Continental Shelf: Ministerial Powers and Controls', 4 (2015) *LSU Journal of Energy Law and Resources*, pp75–95.

[65]   Labour governments were more encouraging of state participation than Conservative governments. Prior to the election of Mrs Thatcher as Prime Minister, the Conservatives were willing to accept some degree of state participation – a policy of tolerance which came to an end with Mrs Thatcher's election.

[66]   Ideological differences on this point are evident in the parliamentary debates that attended the introduction of ss 17 and 18 of the Petroleum and Submarine Pipe-lines Act

context of state participation, this factor led to abrupt switch-backs in British law and policy: most obviously, Labour's creation and the Conservatives' dismantling of the British National Oil Corporation (BNOC). Greater agreement between the main parties means that political ideology has played a less visible role in more recent times. Thatcherism remodelled the British political landscape, with market economics becoming the prevailing creed. Under a centrist leadership, Labour abandoned its commitment to nationalization.[67] Even when Labour entered into a sustained period of power between 1997 and 2010, no attempt was made to re-establish a state national oil company. The election in 2015 of a significantly more left-wing Labour leader, Jeremy Corbyn, raises the possibility of a return to the days of sharp distinctions in the ideological beliefs of the main political parties. But even if this is not so, and if the era of rapid switch-backs based on the temporary ascendancy of a particular political party is over, ideology would not have been rendered wholly irrelevant. The political decision that the interests of the country are best served by a policy of non-intervention is rooted in belief, and the decision to refrain from state participation is as much an ideological choice as is the decision to engage in it.[68] And ideological concerns have arguably played a role in the development of British policy in recent times. For instance, the Conservative-led government's spending cuts significantly reduced the budget of the Department of Energy and Climate Change (DECC) just at the moment when the governance of the UKCS was becoming particularly complicated[69] are likely to have contributed

---

1975, with Labour wishing to reserve the ability to exert control and the Conservatives adopting a more *laissez-faire* attitude.

[67] Clause IV of the Labour Party's Constitution historically contained a commitment to 'the common ownership of the means of production, distribution and exchange'. During Tony Blair's leadership, Clause IV was redrafted and this commitment – generally taken to be the foundation of the party's commitment to public ownership – was removed. See, for example, P Riddell, 'The end of clause IV', 1994–95, 11 (1997) *Contemporary British History* pp24–49.

[68] The point that this fact is sometimes overlooked was forcefully made by the (Labour) Under-Secretary for Energy, John Smith, during parliamentary debate on the Petroleum and Submarine Pipeline Bill 1974–1975: 'One of the things that strikes me as curious about this House is that when the Labour Party brings forward proposals for public enterprise it is regarded as doing that in the cause of ideology, and so it is. We bring forward proposals for public enterprise because we are Socialists. The Conservative Party opposes our proposals root and branch because it is the Conservative Party and believes in private ownership and control, but somehow that is not regarded as ideology of any sort'.

See Hansard, HC Deb 30 April 1975, 5th series, vol 891, col 596.

[69] Throughout the 2000s, life has become progressively more complicated for the regulator. Field size is smaller; there is a greater need for developments to secure access to existing infrastructure; there are more players active within the province and there-

to the light-touch, *laissez-faire* model of governance deployed by DECC. These cuts could be argued to be at least in part driven by an ideological desire for small government, as well as a need to achieve spending cuts due to financial pressures caused by, for example, the banking crisis of 2007–09.

6.2.2.3 Maximizing economic recovery: from political goal to legal obligation

As a result of the Wood Review[70] and its subsequent implementation by government, the expression 'maximizing economic recovery' of petroleum upon the UKCS has considerable currency. We will return to the Wood Review presently. It should, however, first be noted that the notion of maximizing economic recovery has been a factor in UK regulatory practice for some considerable time. The idea of maximizing the value to the State of its oil and gas resources lies at the heart of depletion policy which, as noted above, was a prominent feature of the Labour government's reforms of the 1970s. In the early 2000s, a similar (but not quite identical) idea led to the development of the Fallow Areas and Stewardship initiatives.[71] These were driven not, as depletion policy had been, by the notion of manipulating production levels so as to secure an acceptable financial return to the country, but by the somewhat more straightforward desire to provide government with a means of encouraging oil companies to put the areas that they held under licence to productive use. This desire was in turn driven by concern that if oil or gas from small fields were not produced soon, infrastructure loss could result in the permanent stranding of these deposits; and that while each field loss would in itself be small, the cumulative effect would be large. This concern also drove the gov-

---

fore a greater number of parties to regulate than before; and those players are not just more numerous, but more diverse than before, with late field-life specialists entering the market alongside small start-ups, mid-sized independents and the major players that helped to open up the province.

[70]   Sir Ian Wood, *UKCS Maximising Recovery Review: Final Report*, www.gov.uk/government/uploads/system/uploads/attachment_data/file/471452/UKCS_Maximising_Recovery_Review_FINAL_72pp_locked.pdf ('the Wood Review'), last accessed 18 April 2019.

[71]   Put shortly, the guiding principles behind these schemes is that oil or gas held under licence should be produced promptly where possible, and that the right assets should be in the right hands. Thus with Fallow, the government sought to encourage parties to sell on or surrender assets which they were not actively exploring or producing from. Stewardship was concerned with identifying fields which seemed to be producing sub-optimal quantities of hydrocarbon and subjecting them to particular scrutiny in order to ascertain whether, for instance, disagreements between the licence group were hindering development. For a detailed account of these initiatives, see G Gordon and J Paterson, 'Mature Province Initiatives', in Gordon, Paterson and Üşenmez, op cit n18, Chapter 5.

ernment's repeated, if not wholly successful, attempts to address the problem of third-party access to infrastructure.[72]

Recognition of the significance of the notion of maximizing economic recovery therefore has a long lineage on the UKCS. However the notion was given fresh impetus – and translated from a policy goal to a legal obligation – as a result of the Wood Review on Maximising Economic Recovery.[73] In June 2013 the then Secretary of State for DECC commissioned Sir Ian Wood to undertake a review of the regulation and stewardship of the UK's hydrocarbon reserves. Sir Ian's review concluded that while there was a significant volume of oil still available to be produced from the UKCS, the opportunity to do so would only be fully grasped if – among other things[74] – there were significant changes to the manner in which the UKCS was governed and to the commercial and legal behaviours adopted by the industry. It wished to see an end to the culture of the individual field development and a much more collaborative approach to development, including the adoption of, for example, regional hubs.[75] The review was critical of the under-funding of DECC and of DECC's light-touch mode of regulation.[76] It recommended the creation of a stronger 'arm's-length' regulator, funded directly by an industry levy, which would take greater leadership in the development of strategy and act in cooperation with government and industry in pursuit of the goal of maximizing economic recovery of petroleum on the UKCS as a whole. However, the regulator would be no mere advocate of cooperation and the objective would no longer be a matter of mere policy: licensees and other relevant parties, including infrastructure owners, would be legally obliged to conduct their operations in such a way as to further the objective of maximizing economic recovery, and could be subjected to a range of sanctions if they did not.[77] This vision has duly been brought into law – not, as was initially suggested would be the case, through retroactive changes to the petroleum licence,[78] but through the somewhat convoluted route of passing two pieces of primary legislation which between them impose an obligation to follow the provisions contained in a set of strategy

---

[72]   See, for example, S Rush, 'Access to Infrastructure on the UKCS: The Past, the Present and a Future', Memery Crystal LLP, 2012, pp11–14.

[73]   Op cit n70, above.

[74]   The Review also highlighted other factors, including the need for the industry to embrace new technology.

[75]   Sir Ian was concerned, for instance, to see the number of inefficient floating production-based developments; greater efficiency could be achieved by production using shared facilities: Review, para 2.3.

[76]   Review, para 2.3(iii).

[77]   Sanctions include enforcement notices, financial penalties and, in extreme cases, revocation of the licence: Energy Act 2016 s 42.

[78]   Wood, above n70, at p36.

documents.[79] The new regulator, the OGA, has now begun work.[80] While it is too early to fully assess the effectiveness of the new regime, it clearly ushers in a new era of more interventionist and policy-driven resource management upon the UKCS.

6.3 BRITISH STATE ORGANIZATION AND RESOURCE AND ENVIRONMENTAL MANAGEMENT IN UPSTREAM ACTIVITIES

6.3.1 Seaward Oil and Gas Activities

Throughout the development of the UK oil and gas industry, there has been one key Department ('the relevant department') that has been primarily entrusted with the development and management of upstream oil and gas activities. The identity of the relevant department has changed frequently – sometimes in response to specific policy or geopolitical concerns bearing upon energy, and sometimes as a result of domestic political concerns unrelated to energy matters.[81] This has caused some instability, although this has been less commented upon by the industry than the frequent changes to tax rates and policy. Initially, the relevant department was the Ministry of Power;[82] thereafter, it was the Ministry of Technology,[83] the Department of Trade and Industry (twice)[84]

---

[79] This was achieved by a somewhat convoluted process involving two pieces of primary legislation which between them incorporate into law a provision contained in an OGA strategy document: see Infrastructure Act 2015 s 41 read with the Energy Act 2016 s 42 and OGA, The Maximising Economic Recovery Strategy for the UKCS, available at www.gov.uk/government/uploads/system/uploads/attachment_data/file/509000/MER_UK_Strategy_FINAL.pdf, last accessed 18 April 2019, para 7.

[80] For further information as to the OGA's activities, see its website, www.ogauthority.co.uk/, last accessed 18 April 2019.

[81] The 2016 change, for example, was driven more by the knock-on effects of the need to create a new government department for Brexit than by any specific energy-related concerns.

[82] The Ministry of Power and Fuel, latterly known as the Ministry of Power, was the relevant department until 1969.

[83] The Ministry of Technology, the introduction of which was one of Labour Prime Minister Harold Wilson's flagship initiatives, became the relevant department in 1969. However, the Ministry was abolished by the Conservative Edward Heath when he took power in 1970.

[84] In 1970 the newly elected Conservative government abolished the Ministry for Technology and transferred the upstream oil and gas policy to the Department of Trade and Industry. The oil shock of 1973 led energy to assume greater political significance; as a result, the Department of Energy was formed in 1974. But, having handed responsibility for the upstream oil and gas industry to the Department of Energy, the Department of Trade and Industry became the relevant department again when the Department of

the Department of Energy,[85] and the Department of Business, Enterprise and Regulatory Reform (DBERR),[86] DECC[87] and now the Department of Business, Energy and Industrial Strategy (BEIS).[88]

The precise functions undertaken by the relevant department have changed over time. Historically, the relevant department was not just a licensing authority, entrusted with the task of promoting the development of the upstream oil and gas industry; it was also the industry's principal regulator of safety and environmental matters.[89] The potential for conflict of interest[90] inherent in this position was recognized, at least by some, at a relatively early stage.[91] However, even after the Piper Alpha disaster, the Department of Energy sought to maintain the status quo on the basis that it had accumulated more expertise of the oil industry than other potential regulators. This argument was firmly rejected by the Inquiry into the disaster, which considered that the Department of Energy's approach tended towards 'insularity, over-conservatism, and a lack of ability to look at the regime and themselves in a critical way'.[92] Following the recommendations of the Inquiry,[93] the Health and Safety Executive (HSE) emerged

---

Energy was abolished in 1992. It continued in this role under both Conservative and Labour administrations until 2007.

[85]   The Department of Energy was the relevant department through both Conservative and Labour administrations from 1974 until its abolition by the Conservative government in 1992.

[86]   DBERR was a short-lived department established by Labour Prime Minister Gordon Brown. It existed between 2007 and 2009, but was responsible for upstream oil and gas policy only until 2008.

[87]   DECC was the relevant department from 2008 until 2016.

[88]   The situation is now somewhat complicated by the fact that some of BEIS's functions have been transferred to the OGA.

[89]   J Paterson, 'Health and Safety at Work Offshore', in G Gordon, J Paterson and E Üşenmez, *Oil and Gas Law: Current Practice and Emerging Trends*, 2nd ed, Dundee University Press, 2011, para 8.29.

[90]   See, for example, the evidence provided by the offshore unions in to the Cullen Inquiry: Lord Cullen, The Public Inquiry into the Piper Alpha Disaster, 1990, Cm 1310, Voll II, para 22.13. A similar conflict was also recognized in the USA following the Deepwater Horizon disaster, when the Minerals and Management Service, which had also acted as the industry's sponsor and regulator, was split up by executive order: see G Gordon, 'The Deepwater Horizon Disaster: The Regulatory Response in the United Kingdom and Europe', in R Caddell and R Thomas, *Shipping, Law and the Marine Environment in the 21st Century*, Lawtext, 2013, at p183.

[91]   See, for example, the Note of Dissent to the Burgoyne Report, referred to by Paterson, n32 above, at 8.24.

[92]   Lord Cullen, 22.20.

[93]   Cullen, Recommendations 23–26, pp391–392.

as the sole regulator of offshore health and safety.[94] However, the Inquiry was concerned with offshore safety, not environmental regulation. The relevant department continued to have responsibility for environmental regulation, a position which BEIS continues to hold.[95] Arguably, although environmental matters were dealt with by a separate specialist team within the department, this dual role presented the potential for conflict; however, the OGA's assumption of the State's resource management role would seem to remove this risk. Statutory consultancies (eg, the Joint Nature Conservancy Committee (JNCC), with which the OGA must consult before making acreage available for licensing purposes) could also be said to have an indirect but significant role in the supervision and enforcement of law and regulation in this area.[96] BEIS as environmental regulator[97] works alongside the HSE as the Offshore Safety Directive Regulator,[98] the Competent Authority for the purposes of implementing the Offshore Safety Directive's provisions on prevention and/or limiting the consequences of major offshore accidents.[99]

Fiscal (taxation) policy is determined not by BEIS, but by the Treasury which, as we shall see below, has frequently increased the rate of taxation.[100] Sometimes this has happened abruptly and without consultation, leading to allegations that the pursuit of short-term gain has led it to engage in behaviours that destabilize long-term investor confidence.[101] More recently, however –

---

[94] As regulator, the HSE's function is twofold. It responds to accidents (or 'near-misses') by investigating the circumstances of such events and issuing enforcement notices or prosecutes parties in breach of the relevant regulations. It also has a more proactive, agenda-setting role, in that it may, through its Key Programme system, identify strategic priorities and liaise with the industry in order to improve practice in specific areas of operation. For more on the Key Programme initiatives, see www.hse.gov.uk/offshore/programmereports.htm, last accessed 18 April 2019.

[95] Guidance on BEIS's environmental policies can be found at DECC, *Oil and Gas: Environmental Policy*, available at www.gov.uk/guidance/oil-and-gas-environmental -policy, last accessed 18 April 2019.

[96] For more on the JNCC's role, see JNCC, Oil and Gas, http://jncc.defra.gov.uk/ page-4275, last accessed 18 April 2019).

[97] BEIS has also retained responsibility for decommissioning.

[98] For more information, see www.hse.gov.uk/osdr/authority/index.htm, last accessed 18 April 2019.

[99] For an account of the Offshore Safety Directive, see G Gordon, 'Offshore safety: the European Commission's legislative initiatives' in M Roggenkamp and H Bjornebye (eds), *European Energy & Law Report X*, Intersentia, (2014), pp139–172.

[100] See further below.

[101] UK Offshore Operators Association, *Response to Budget Statement*, 17 April 2002, http://oilandgasuk.co.uk/ukooa-response-to-the-budget-statement/, last accessed 19 April 2019; HC Deb 03 July 2002 at Col 279 per A Salmond MP, accusing the government of making a fundamental error in undertaking 'a short-term, smash-and-grab

most obviously following the sudden price drop of 2014,[102] although the trend was evident even before then[103] – the Treasury has shown a much increased awareness of the need to use fiscal policy as a means of encouraging investment. While rapid changes have been made to the taxation regime over the last few years, these have been in an attempt to address the problems caused by low oil price and have been welcomed by the industry.[104]

Government – in the form of BEIS – continues to have the primary responsibility for policy formulation; albeit that, following the implementation of the recommendations of the Wood Review on Maximising Economic Recovery,[105] much of the day-to-day responsibility for licensing, governance and stewardship of the industry has been transferred to the OGA.[106]

The relationship between BEIS and the OGA is not altogether straightforward and requires further consideration. Although the Wood Review conceived of the OGA as an 'arm's-length' regulator,[107] the OGA is by no means wholly independent of BEIS. Initially established, for reasons of speed and expediency, as an Executive Agency of DECC, the OGA became a Government Company on 12 July 2016.[108] As the Energy Bill was passing through Committee stage, Energy Minister Andrea Leadsom stated that: 'Classification as a Government company will enable the OGA to have operational independence from Government and will provide a more suitable

---

raid to maximise revenue in the next few years'; Scottish Government, White Paper: *Scotland's Future* (2013), p300.

[102] See, for example, HM Treasury, Driving Investment: A Plan to Reform the Oil and Gas Fiscal Regime, 2014, www.gov.uk/government/uploads/system/uploads/attachment_data/file/382785/PU1721_Driving_investment_-_a_plan_to_reform_the _oil_and_gas_fiscal_regime.pdf, last accessed 18 April 2019.

[103] See, for instance, the Treasury's willingness to enter into Decommissioning Relief Deeds in order to provide the industry with stability on the question of the available level of tax relief against decommissioning expenses: www.parliament.uk/documents/commons-vote-office/July%202015/21%20July/6-Treasury-Decommissioning-relief .pdf, last accessed 18 April 2019.

[104] Oil and Gas UK, 18 March 2015, *Budget lays strong foundations for regeneration of the UK North Sea*, http://oilandgasuk.co.uk/budget-lays-strong-foundations-for -regeneration-of-the-uk-north-sea/, last accessed 19 April 2019.

[105] The Wood Review called for a new, better-resourced and more powerful regulator of the oil and gas industry. See further the discussion at nn76–77 and associated text, above.

[106] Energy Act 2016 s 2 and Sched 1. It should not be thought, however, that the OGA has operational responsibility for all matters. Operational responsibility for environmental regulation and decommissioning activities remains with BEIS.

[107] Sir Ian Wood, *UKCS Maximising Recovery Review: Final Report*, op cit. n70, p21.

[108] The Energy Act 2016 (Commencement No. 1 and Savings Provisions) Regulations 2016 (SI2016/602) Reg 3.

platform and the regulatory certainty that the industry requires to invest in exploration and production activity.'[109]

It may well be that this structure will provide some de facto independence;[110] but as the Framework Document which governs the relationship between BEIS and the OGA makes clear, BEIS continues to have ultimate responsibility for the development of policy, and the OGA is obliged to follow BEIS's policy lead.[111] BEIS is statutorily empowered to issue directions to the OGA 'where the Secretary of State considers that the directions— are necessary in the interests of national security, or are otherwise in the public interest'.[112]

In 'exceptional circumstances', the power to direct extends to 'the exercise of a regulatory function in a particular case'.[113] Moreover, the Secretary of State for BEIS is the sole shareholder of the OGA,[114] and the Framework Document makes it clear that the OGA's operational independence is subject to the 'rights the Secretary of State has by virtue of being the sole shareholder of the OGA'.[115] It also recognizes BEIS's right to create or amend regulations, and that these could impact upon the OGA's operations; and provides that there is a range of areas where the OGA requires BEIS consent before acting. Taking all of this together, although the OGA is, in practice, acting as licensing authority and resource manager on a day-to-day basis, BEIS has a great many tools at its disposal to exert control over the OGA.

6.3.2   Landward Oil and Gas Operations

In the onshore area, subject to what is said about devolution below, all of the parties mentioned above are involved in resource management and environmental and other regulatory activities. Onshore, however, the regulatory

---

[109] Public Bill Committee, Energy Bill (First Sitting), Tuesday 26 January 2016, cols 4–5.

[110] See also BEIS and OGA, *Oil and Gas Authority Framework Document*, 2016, available at www.ogauthority.co.uk/media/3154/oga-framework-document-december -2016.docx, last accessed 19 April 2019, para 6.

[111] DECC and OGA, *Oil and Gas Authority Framework Document*, 2015, available at www.gov.uk/government/uploads/system/uploads/attachment_data/file/420028/ OGA_Framework_Document_April_2015.pdf, last accessed 19 April 2019. Among the purposes and duties of the OGA listed on p9 is 4(D), 'to adhere to and implement Government policy set by the Secretary of State'.

[112] Energy Act 2016 s 9(1).

[113] Energy Act 2016 s 9(2).

[114] DECC, *Government Response to Sir Ian Wood's UKCS Maximising Economic Recovery Review*, 2014, available at www.gov.uk/government/uploads/system/uploads/ attachment_data/file/330927/Wood_Review_Government_Response_Final.pdf,   last accessed 19 April 2019, para 2.12.

[115] Framework Document, para 7(D).

framework is yet more complex, because of the need to obtain planning consents and satisfy further environmental standards.[116] Devolution adds a further level of complexity. The Northern Irish government is the licensing authority for onshore oil and gas operations in Northern Ireland. Historically, onshore licensing throughout the rest of the UK was administered by the relevant department of the UK government; however, following the recommendations of the Smith Commission on Further Devolution to Scotland, the Scottish government has become the licensing authority for oil and gas operations within the Scottish onshore area.[117] Although the policy of the devolved assemblies is, at least at present, not to permit unconventional onshore operations,[118] the UK government supports the development of a shale gas industry. It has stated that it wishes to streamline the onshore system, the complexity of which it recognizes may be acting as a barrier to the development of unconventional gas, and some progress has been made in this regard[119] However, onshore development – in particular, fracking – continues to be highly contentious due to a combination of climate change and more localized environmental concerns, and thus far the government has had only limited success in its attempts to kick-start this industry.[120]

6.4 BRITISH GOVERNMENTAL PARTICIPATION IN UPSTREAM ACTIVITIES

As has already been noted, the legal framework for offshore oil and gas operations emerged in the mid-1960s, when s 1(3) of the Continental Shelf Act 1964 exported the substantially untested onshore petroleum regime into

---

[116] See, for example, G Gordon, A McHarg and J Paterson, 'Energy Law in the UK', in M Roggenkamp, C Redgwell, A Ronne and I Del Guayo, *Energy Law in Europe*, 3rd ed, 2016, at 14.31–14.35; J Paterson and T Hunter, 'Shale Gas Law and Regulation in the United Kingdom' in T Hunter (ed) *Handbook of Shale Gas Law and Policy*, Intersentia, 2016, Chapter 13.

[117] Scotland Act 2016 s 49(1).

[118] Utility Week, *Northern Ireland bans fracking* 20 September 2015; The Scottish Government, Press Release: *Moratorium Called on Fracking*, 28 January 2015.

[119] For instance, the statutory regime contained in ss 43–48 of the Infrastructure Act 2015 provides a clear right to deep-level access for petroleum operations, removing a potentially significant barrier to onshore development: see DECC, Press Release – *Government proposals to simplify deep underground access for shale gas and geothermal industries*, 23 May 2014.

[120] G Gordon, A McHarg and J Paterson, 'Energy Law in the UK', in M Roggenkamp, C Redgwell, A Ronne and I Del Guayo, *Energy Law in Europe*, 3rd ed, 2016, at 14.34.

the offshore area.[121] In the licences granted under this regime, state companies participated alongside private investors, but initially in a rather piecemeal and ad hoc way.[122] One would search the relevant petroleum legislation in vain for any authoritative statement of the intended relationship between industry and State, but given that the UK system operated by means of the grant of a broad discretionary power to the licensing authority, no such legislative provision was necessary;[123] instead, decisions on the extent of state participation were announced on a round-by-round basis. These decisions reflected the state of government knowledge and the government's attitude to risk, as well as the political ideology of the government of the day. Broadly speaking, the rounds taking place under the aegis of the Labour governments of 1964–70 and 1974–79 involved more public participation than those under the Conservative administrations of 1959–64 and 1970–74. Political ideology was, however, of only limited relevance prior to 1974; throughout the period of so-called post-war consensus politics that prevailed in Britain in the 1960s and early 1970s, even Conservative governments were willing to accept at least some degree of direct state participation in the oil and gas industry,[124] and some senior Conservative politicians actively advocated the creation of a state oil company.[125] Moreover, prior to 1974, Labour's ambitions for state participation were tempered by an acute awareness of the risk of oil and gas operations.[126] A brief discussion of the history of the first few licensing rounds will help to illustrate these points.

---

[121] The onshore petroleum regime was contained in the Petroleum (Production) Act 1934. Although the Act had been in force for almost three decades, very few commercial finds of oil or gas had been made onshore. The government therefore had little prior experience of the oil and gas industry prior to the development of the offshore area. See further G Gordon, 'Petroleum Licensing', in G Gordon, J Paterson and E Üşenmez, *Oil and Gas Law: Current Issues and Emerging Trends,* 2nd ed, 2011, para 4.8.

[122] See T Daintith, 'Appendix A – State Participation' in Daintith, Willoughby and Hill ('Daintith, "State Participation"') para 1–A03.

[123] For more on the discretionary nature of the licence, see Gordon, 'Petroleum Licensing', in Gordon, Paterson and Üşenmez, op cit n18, para 4.14.

[124] One Nation Conservatism, the dominant strand of Conservative thought in the post-war era, had at its core the notion that the State could be a promoter of social cohesion, and was tolerant of at least some level of state participation in the UK's industrial sector: see, for example, H Bochel, 'One Nation Conservatism and Social Policy, 1951–64', (2010) 18 *Journal of Poverty and Social Justice,* 123; P Dorey and M Garnett, 'The Weaker-willed, the craven-hearted: the decline of One Nation Conservatism', (2015) 5 *Global Discourse,* 69.

[125] In 1973, Secretary of State for Trade and Industry Peter Walker favoured the creation of a new state oil company: Kemp, *Official History: Vol 1,* n37, above, p264.

[126] Ibid, p111. Peter Odell, the great oil economist, lasted only a year as an adviser to the Labour Party before resigning in frustration at its lack of enthusiasm for his radical vision for the development of oil and gas assets upon the UKCS.

In the first licensing round, held under the auspices of a Conservative government in 1964, no preference was extended to public sector British industry; however, the Gas Council[127] participated 'on equal footing with private enterprise'[128] and secured a modest proportion of the interests on offer.[129] In the run-up to the second round, held in 1965, the now-Labour government gave consideration to a range of options intended to boost both the participation of indigenous private sector companies and the State. For a variety of reasons – including industry opposition,[130] worries about accepting excessive exploration risk[131] and concerns as to the technical capacity of the public sector actors[132] – the government ultimately decided against the more radical proposals under discussion, including the formation of a state national oil company.[133] It instead opted to provide some modest encouragement for state participation by announcing in Parliament that, in the event of competition for acreage, a preference would be granted to any consortium willing to participate with the National Coal Board (NCB) or British Gas Corporation (BCG).[134] It considered that larger-scale state participation should wait until at least the third round in 1970, by which time licence relinquishment provisions would mean that a significant quantity of acreage was due to be re-let;[135] it was hoped that by then, the government would be able to 'properly evaluate the risks' and 'would be able to allocate acreage to nationalised industry with a fairly full knowledge of what was involved'.[136] By the time of the third round, however,

---

[127] The Gas Council, the forerunner of the British Gas Corporation, was initially conceived of as a point of liaison between central government and the regional Gas Boards responsible for the distribution of gas within their locality.

[128] Daintith, 'State Participation', n122, above, 1–A03.

[129] Overall, the Gas Council was allocated a 3 per cent equity share in the licences offered under the first round: Kemp, *Official History: Vol 1*, p60. British Petroleum also participated and was awarded a 6.5 per cent overall share. However, although the UK government at the time held a significant shareholding in BP, a long-standing policy of non-intervention in that company's commercial affairs means that it is not appropriate to view that company as a public corporation in a strict sense: Daintith, 'State Participation', n122, above, 1–A03. In 1966, the statutory purposes of the National Coal Board were amended to permit it to additionally participate in the offshore oil and gas industry: National Coal Board (Additional Powers) Act 1966 s 1.

[130] Kemp, *Official History: Vol 1*, n37, above, p92.

[131] Ibid, p91. This risk could have been mitigated by the inclusion of a carried interest provision, but this appears not to have been seriously considered.

[132] Ibid, p92.

[133] Another reason why it decided not to form a state national oil company at this time was the concern that doing so could be a source of embarrassment for existing British oil companies in their international operations: ibid.

[134] HC Deb Vol 721 cols 507–512 (24 November 1965).

[135] Kemp, *Official History: Vol 1*, n37, above, p111.

[136] Ibid, p97.

the government's ambitions for participation had been somewhat scaled back. The idea of launching a National Hydrocarbons Corporation had been mooted by the Labour Party's North Sea Study Group as early as 1967 'as a means of building up a public stake in North Sea oil without the expense of full nation-alisation'.[137] This proposal was, however, rejected at Ministerial level on the basis that the economic benefits of the venture were uncertain.[138] Instead, the government sought to promote the participation of state-owned enterprise by taking the incremental step of making a willingness to permit participation by the NCB or BGC not merely desirable, but 'an essential prerequisite for the grant of a licence'.[139] By the time of the fourth round – under which a very large acreage was let, and which Kemp describes as 'momentous'[140] – Labour had lost power to the Conservatives, which fell back to the same policy that they had adopted at the time of the first round: public participation was per-mitted, but not made compulsory, and the details of such participation was left as a matter between the public corporations and the private sector entities with which they contracted.[141]

Thus, up to this point, while Labour and the Conservatives differed in the extent to which they actively encouraged state participation, both were to some extent open to the notion. However, North Sea oil was set to become a highly politicized issue. A combination of the criticism of tax policy contained in the Public Accounts Committee Report of 1972–73 and the sharp increase in oil price caused by the oil shock of 1973 – as well as a growing awareness of the scale of the deposits to be found in the North Sea – led to an intense focus upon oil and gas policy.[142] As we have seen, this led to the creation of the Department of Energy. It also prompted Labour to formulate policies that would appeal to the voter concerned about 'dependence for oil on OPEC and the perception of "excess" profits accruing to, mainly foreign-owned, oil multinationals'.[143] This led Labour to renew its interest in state control. When returned to office in 1974, Labour set about introducing 'systematic state participation as a means of redressing. . . inadequate protection of the national interest'.[144] BNOC was

---

[137] Ibid, p111.

[138] Ibid.

[139] Daintith, 'State Participation', n122, above, 1–A03.

[140] Kemp, *Official History: Vol 1*, n37, above, p247.

[141] Daintith, 'State Participation', n112, above, 1–A03.

[142] D Hann, 'The process of government and UK oil participation policy', 1986 *Energy Policy* 253–261 ('Hann, "UK oil participation policy"') p254.

[143] Ibid, p254.

[144] Daintith, 'State Participation', n122, above, 1–A04. The proposals are set out in detail in HM Government, *UK Offshore Oil and Gas Policy*, Cmnd 5696, HMSO, London, 1974.

established by the Petroleum and Submarine Pipe-lines Act 1975.[145] BNOC set about securing upstream oil and gas interests through a number of different means. It acquired, on arm's-length commercial terms, the business of some commercial entities[146] and entered into participation agreements with most of the licensees then active on the UKCS relative to previously licensed interests.[147] Moreover, in the fifth licensing round, although no formal change was made to the legislative framework under which the licences were offered,[148] when announcing the round, the government stipulated that all successful applicants must enter into a satisfactory arrangement whereby BNOC or BGC would take a 51 per cent interest in the licence. Under the sixth round of 1978–79, the 51 per cent interest requirement was retained, and BNOC and BGC's position was further enhanced by the government's inclusion of four further criteria to which it would have regard in deciding whether to grant the licence:

- the extent to which the applicant was prepared to carry the State's interest in the licence;
- the extent to which the applicant was prepared to offer BNOC or BGC an equity share of greater than 51 per cent;
- whether the applicant was willing to offer BNOC a right of first refusal on the applicant's share of produced hydrocarbons; and
- whether the applicant was willing to grant BNOC the option to sell its share of produced hydrocarbon to the applicant.[149]

By the end of the sixth round, BNOC and BGC had secured for the State a significant equity interest in offshore licences.[150] This situation was not, however, destined to last for long. The general public's appetite for state ownership of utilities and industry was diminished by the fractious relationship between trade unions and the State throughout the 1970s.[151] In 1979, Labour

---

[145] The Petroleum and Submarine Pipe-lines Act 1975, Part I.

[146] For example, Burmah Oil. See Kemp, *Official History: Vol 1*, n37, above, pp 425–26.

[147] Daintith, 'State Participation', n112, above, 1–A10. It should be noted that under these participation agreements, although BNOC became entitled to a significant share of oil at market price, its voting rights were circumscribed and it did not gain an equity share in the licensed interests. For a listing of all the companies with which such agreements were signed, see Kemp, *Official History: Vol 1*, n37 above, p438.

[148] The inherent flexibility of the discretionary nature of the UK's licence system meant that no such changes were necessary.

[149] *London Gazette*, 8 August 1978, pp9507–10. See in particular paras 5(k)–(n).

[150] Nelsen, *Petroleum Policy in Britain and Norway*, n52, above, pp320–321.

[151] This fractious relationship was in turn fed by a range of socio-economic factors, including job losses caused by globalization's effect upon the UK's industrial base,

lost power to a revitalized Conservative party led by Margaret Thatcher. Thatcher was not a One Nation Conservative, willing to tolerate a consensus policy of state participation. Thatcherism involved free market economics and a radically reduced role for the State;[152] and, ideological issues apart, the need to reduce the Public Sector Borrowing Requirement provided a significant practical impulse to secure funds by privatizing state assets.[153] Among the first state-owned undertakings to be dismantled were BNOC (privatized in stages between 1982 and 1985)[154] and BGC, which was privatized in 1986.[155]

Since the privatizations of the mid-1980s, there has been no direct participation in the upstream oil and gas industry in Great Britain or upon the UKCS. As we have seen, between the mid-1980s until almost the present day, the State's interest has been limited to that of licensor and steward, and generally a fairly hands-off steward at that.[156] The implementation of the Wood Review changes things – not to the extent of recreating a state national oil company, but at least to the extent of creating a new and more active mode of governance for the sector.[157] The OGA is in the process of formulating a range of strategies;[158] it has the power to attend JOA[159] meetings and to impose sanctions for breaches of petroleum-related requirements;[160] it is at the heart of trying to foster a more

---

wage demands fuelled by the fall in sterling and the drive towards equality in the workplace. For a comprehensive account, see, for example, T Martin Lopez, 'The Winter of Discontent: Causes and Context', in T Martin Lopez and S Rowbotham, *The Winter of Discontent: Myth, Memory, and History*, 2014, Liverpool University Press, Chapter 2.

[152] See, for example, P Jackson, 'Thatcherism and the Public Sector', (2014) 45 *Industrial Relations Journal*, 266.

[153] Kemp, *Official History: Vol 1*, n37, above, pp489–499.

[154] Daintith, 'State Participation' n122, above, paras 1–A15-1–A20.

[155] British Gas was privatized – very ineptly – by the Gas Act 1986. For an account of the run-up to the privatization, and the subsequent aftermath, see A Kemp, *Official History: Vol 2: Moderating the State's Role* (2011, Routledge) Chapter 1.

[156] See section 6.2, above.

[157] Although not created for this purpose, it is not impossible to conceive of a situation where, if oil price remains low and insolvencies threaten the viability of the sector, the OGA's role evolves into that of an operator of last resort. However, this is pure conjecture.

[158] It has, for instance, recently published its Decommissioning Strategy: OGA, Decommissioning Strategy, 2016, available at www.gov.uk/government/uploads/system/uploads/attachment_data/file/533973/OGA_DECOMM_v1.1.PDF,   last accessed 19 April 2019.

[159] Energy Act 2016, Chapter 4.

[160] Energy Act 2016, Chapter 5. 'Petroleum-related requirement' is defined by s 42(3) to include breaches of licence terms and conditions and breaching the statutory obligation to follow the OGA's strategy on maximizing economic recovery.

collaborative industry mindset and is championing the use of technology.[161] Its attempts to encourage exploration in frontier areas go beyond the innovative licensing models introduced by DECC in the mid-2000s, and extend to the commissioning of seismic surveys in frontier areas for the use of the industry at large.[162] This last initiative provides an example of how recognizing that the State's legitimate interest in the oil and gas sector goes beyond the collection of tax revenue could potentially result in the unlocking of value which might otherwise never have been realized.

Does the creation of the OGA mark the first stage of a return to direct state participation in the oil industry? The OGA was not conceived of as a state national oil company, and at time of writing, it does not seem likely that it will become one in the near future. However, it is possible to conceive of a set of circumstances where a situation akin to the banking crisis of the late 2000s leads to the need for an at least temporary state intervention in the industry. It is, for example, not unthinkable that a lack of cash flow caused by a period of low oil price could lead to a number of oil companies urgently needing to sell assets at a time when buyers are absent from the market. This, in turn, could lead to insolvencies and the risk of contagion throughout the sector. One means of addressing such a problem could be for the OGA to end up acting as an operator of last resort. It is to be hoped that, if Britain does ever revert to direct participation, it does not happen in such desperate circumstances.

6.5 BRITISH LOCAL CONTENT AND SOCIAL FUNDS

6.5.1 Local Content

'Local content' has been defined as 'the local resources a project or business utilizes or develops along its value chain while invested in a host country'.[163] It thus refers not just to direct participation interests in upstream investments, but also to the supply chain. The appeal of a policy encouraging local content

---

[161] See, for example, the OGA's attempt to bring operators and the supply chain together to uncover cost efficiencies and innovative technological solutions applicable to the Southern North Sea's well plug and abandonment market: OGA, *OGA Leads Southern North Sea Hackathon*, 23 June 2016, available at www.gov.uk/government/news/oga-leads-southern-north-sea-hackathon, last accessed 19 April 2019.

[162] See further OGA, *Oil and Gas: 2015 UK Government-funded seismic project*, available at www.gov.uk/guidance/oil-and-gas-uk-government-funded-seismic -project, last accessed 19 April 2019.

[163] IPIECA, *Local Content: A Guidance Document for the Oil and Gas Industry*, 2nd ed, 2016, www.ipieca.org/sites/default/files/publications/Local_Content_2016 .pdf, last accessed 19 April 2019, at 8.

is obvious: all other things being equal, the greater the proportion of project money spent within the country, the greater the benefit to the national economy accruing from the development. Promoting local content is therefore often seen by policy makers as a means of maximizing the benefit accruing to the State as a result of exploiting petroleum deposits.[164] Thoughtfully designed local content policies can have beneficial effects, increasing the value added to the economy by the industry, addressing specific market failures, and supporting the host government's social policy objectives.[165] However, poorly designed local content policies can have a serious distorting effect within the market,[166] creating protected industrial sectors with little incentive to avoid inefficiency and waste. As well as failing to maximize the opportunity to upskill the local workforce, these factors can lead to a lack of innovation, project delays and increased costs; factors could have an adverse impact upon investment decisions[167] and even upon design and safety standards. Local content policies are also recognized as providing opportunities for corrupt practices.[168] Thus, the promotion of local content is not an unequivocal good. Great care must be taken in the design of such policies.

Some petroleum states pursue rigid local content policies which stipulate a firm minimum level of local content in petroleum activities. The detail of the rules implementing these policies varies from state to state. Some stipulate firm local participation rules, providing for a minimum percentage interest that must be held by indigenous companies[169] or a requirement for a particular project's workforce to contain a certain proportion of the host state's nationals.[170] Other local content provisions may be more nuanced and process driven, requiring a bidder to provide information about the extent to which its pro-

---

[164]   S Tordo, M Warner, O Manzano, Y Anouti, *Local content in the oil and gas sector – A World Bank study*, 2013, World Bank, available at http://documents.worldbank .org/curated/en/2013/01/17997330/local-content-oil-gas-sector, last accessed 19 April 2019, 1.

[165]   Ibid pp24–26.

[166]   Ibid p159.

[167]   In an extreme situation, production within a particular province could become uneconomical as a result of the additional costs arising from a requirement to contract with a highly inefficient local service sector.

[168]   These risks are discussed, and a risk-mitigation strategy sketched, in IPIECA, n163, above, pp33–36.

[169]   See, for example, the Ghanaian Petroleum (Local Content and Local Participation) Regulations 2013, LI 2204, which, among other things, provide for a minimum of 5 per cent equity participation by an indigenous Ghanaian company (other than Ghanaian National Petroleum Company) in all petroleum licences.

[170]   See, for example, the Nigerian Oil and Gas Industry Content Development Act, 2010, which, among other things, provides that no more than 5 per cent of management positions may be filled by expatriates: Art 32.

posed project will benefit the local economy;[171] while no minimum threshold is stipulated, the bidder will of course understand that, all things being equal, the host state will prefer to see more local content than less.

At present, the UK has no specific local participation or local content rules.[172] Historically, however, this has not always been the case. As we have already seen, three licensing rounds were undertaken in circumstances where direct state participation was mandatory.[173] State participation, although not formally required, was also encouraged in the second round, and the government was also concerned to secure good levels of domestic private sector investment in the offshore oil and gas industry.

Outside the specific context of licence participation, the UK government took other measures to try to foster local content. Initially, the UK government's policy was directed towards 'the rapid and thorough development of the United Kingdom Continental Shelf and on the reduction of our dependence on foreign oil'.[174] Relatively little government attention was paid to the supply sector at this time; however, as the scale of North Sea oil deposits began to become clear, so too did the realization that this represented a significant prize for the supply sector – one that was likely to be largely lost overseas if

---

[171] See, for example, the requirement under the Ghanaian regulations to submit quarterly and annual local content plans.

[172] Some policies to some extent resemble local content policies, but strictly speaking are not. For instance, under the aegis of MER UK, the OGA is attempting to encourage cost reduction and improved efficiency within the supply chain, and to encourage innovation and the uptake of new technological solutions: see, for example, OGA *Supply Chain Strategy* (2016), available at www.ogauthority.co.uk/supply-chain/strategy/; OGA, *Technology Strategy* (2016), available at www.ogauthority.co.uk/news-publications/publications/2016/technology-strategy/, last accessed 19 April 2019. If successful, these strategies may help local companies to become more competitive locally and indeed internationally, and if so, this may enhance the success of UK-based service sector companies. However, that is not the primary purpose of the strategies: they are directed towards the OGA's objective of maximizing economic recovery of petroleum from the UK as a whole.

[173] That is, the third, fifth and sixth rounds. Dam argues that the very structure of the system – in particular, its discretionary nature – had led to a higher degree of participation by British companies than would have been possible with a cash premium bidding system: Dam, n49 above, p40.

[174] *Arrangements for the Exploitation of Petroleum and Natural Gas within Great Britain and the Continental Shelf, Memorandum by Department of Trade and Industry, First Report from the Committee of Public Accounts*, Session 1972–73, North Sea Oil and Gas, HC 122, HMSO, 1973, p33. See also N Smith, *The Sea of Lost Opportunity: North Sea Oil and Gas, British Industry and the Offshore Supplies Office.* (2011) Amsterdam: Elsevier, p97. For more on Britain's parlous financial situation at the time, and in particular its balance of payments problem, see Smith, p20.

action was not taken.[175] By the early 1970s, a policy of supporting the local supply industry had been adopted.[176] Central to this policy was the notion that the industry should provide a 'full and fair opportunity' to compete for orders.[177] This was implemented through the work of the Offshore Supplies Office, a division of the Department of Energy formed in 1973, fortified by the Department of Energy incorporating the 'full and fair opportunity' criterion into its licence allocation procedure. The notice in the *London Gazette* which announced the sixth licensing round demonstrates this: the government stated that applications would be 'judged against the background of the continuing need for expeditious, thorough and efficient exploration to identify oil and gas resources of the U.K. Continental Shelf';[178] but among the factors which would be particularly borne in mind in examining applications was:

> whether the applicant subscribes to the Memorandum of Understanding agreed by the Secretary of State and United Kingdom Offshore Operators Association to ensure that full and fair opportunity is provided to U.K. industry to compete for orders of goods and services. Where the applicant is or has been a licensee, his past performance in providing full and fair opportunity to U.K. industry will be taken fully into account.[179]

This policy – although towards the softer end of the spectrum of local content measures sketched above – at first blush provided quite a significant incentive for an oil company to sign up to the Memorandum and then adhere to its terms, and the Offshore Supplies Office (OSO) made concerted efforts to promote the development of a domestic supply sector, including proactive steps such as encouraging joint ventures between UK and foreign companies in order to address capability gaps.[180] Ultimately, however, before the entry into force of the Hydrocarbons Licensing Directive 1994[181] meant that the policy had to

---

[175] Smith, p98.

[176] A study conducted by Smith identifies 'Encouraging the British Supply Industry' as the fourth most significant consideration among policymakers active during the 1970s: Smith, above n174, p79.

[177] Kemp, 'An assessment of UK North Sea oil and gas policies: Twenty-five years on', 1990 *Energy Policy* 599, p602.

[178] *London Gazette*, Issue 47612, 8 August 1978, p9508, para 5.

[179] Ibid, para 5 (h).

[180] Smith, n174 above, p80.

[181] Directive 94/22/EC. This provided a set of non-discriminatory criteria which the Minister was bound to utilize when determining licence applications: for instance, the Directive provided that it was legitimate for the Minister to have regard to the technical and financial capability of the applicant and any prior lack of efficiency or responsibility in the execution of operations under a petroleum licence: Reg 3(1). The local content provisions described above were discontinued from the 16th licensing round onwards.

be discontinued as potentially discriminatory and protectionist,[182] the OSO's efforts met with limited success. Smith, a former Director of OSO, notes that while the UK achieved some success in fields of activity requiring labour rather than technology, or which had a strong locational factor, 'few indigenous offshore service and supply businesses of truly international scale developed'.[183] Among other reasons, he considers that the government's focus on balance of trade issues meant that the importance of the supply sector became apparent to the government too late; that, although mindful of the importance of the supply sector, developing it was never the government's primary policy; that poor performance by some UK contractors in early projects leading to a lack of confidence on the part of the industry, as well as a lack of interest on the part of some other contractors reluctant to commit to an industry which they thought might prove to be a flash in the pan; that the UK was hampered by a lack of venture capital, and that in any event, by the time the North Sea came onstream, a very significant technological advantage had been opened up by the foreign supply sector.[184]

At time of writing, it seems likely that the UK will ultimately leave the EU following the referendum in 2016. The government has announced that the general strategy of the Great Repeal Bill which will follow Brexit will be 'to convert EU law as it applies in the UK into domestic law on the day we leave'.[185] Thus, while in the longer term, the UK's departure from the EU opens up the possibility of a change in policy, in the immediate short term at least, it would seem that the UK will continue to operate in accordance with the precepts of the Hydrocarbons Licensing Directive.

6.5.2 Social Funds

Unlike many other petroleum-producing states,[186] the UK has never sought to stand out or hypothecate any part of the State's take from oil and gas pro-

---

[182] Even before the Hydrocarbons Licensing Directive, the policy of encouraging local content was of dubious legality in EU law. European law and General Agreement on Tariffs and Trade obligations played a role in the decision to word the UK's local content provisions requirements in a non-prescriptive manner: A Kemp, *Official History: Vol 1* p254. Nevertheless, the European Commission warned the UK as early as 1985 that these criteria might be contrary to European law: Daintith, Willoughby and Hill para 1–317.

[183] Smith, n174 above, pxvi and p216.

[184] Smith, n174 above, pp247–253.

[185] Department of Exiting the European Union, White Paper: Legislating for the United Kingdom's withdrawal from the European Union, Cmnd 9446, (2017), p7.

[186] Norway is one of the leading examples of such a state. Many other examples could be given: see, for example, the list provided in the Scottish Government, *An Oil*

duction. It has neither a sovereign wealth fund[187] nor a stabilization fund.[188] The revenue generated by taxing the profits on oil and gas activities – which has been very considerable[189] – has flowed to the exchequer and been applied for general purposes. None of it has been put aside to fund the State's very considerable share of decommissioning costs, far less a contingency fund to cover against, for instance, the risk that some licensees may be unable to meet their part of the decommissioning costs or the risk that in the long term, some liabilities may fall upon the State.[190]

It does not appear to have led to an increase in expenditure on infrastructure projects[191] and neither does it appear to be true to say that significant volumes of the revenue from North Sea oil were spent on welfare payments while Britain moved from a manufacturing to service-based economy during the 1980s and 1990s.[192] Instead, the majority of the oil tax money was used to reduce the rate of general taxation.[193] It could therefore be argued that part of the nation's capital stock has been used up during the oil-producing era to the

---

Fund for Scotland, 2009, available at www.gov.scot/Resource/Doc/280368/0084457 .pdf, last accessed 19 April 2019, Table 9, p35.

[187] Sometimes called a social fund – that is, a fund into which a share of oil monies is paid in order to maintain a set proportion of the nation's capital stock for the benefit of future generations. For an excellent account of the benefits and optimal means of managing such funds, see A Cummine, *Citizen's Wealth: Why (and How) Sovereign Funds Should Be Managed By the People for the People*, Yale University Press, 2017.

[188] That is, a fund into which cash is paid during times of unusually high oil price and from which money may be drawn down in order to maintain public spending levels during times of low oil price.

[189] Total tax take since offshore oil and gas production began has been a little under £190 billion: author's calculation using data from HM Revenue and Customs, *Government Revenues from UK Oil and Gas Production 1968/9–2015/6*, available at www.gov.uk/government/uploads/system/uploads/attachment_data/file/532649/Table _11_11.pdf, last accessed 19 April 2019.

[190] Apart from the State's support of decommissioning via the tax system (discussed further below), it is British government policy that the cost of decommissioning should be borne by the oil companies that had the benefit of production from the relevant field.

[191] Public sector net investment fell from 2.5 per cent of gross domestic product in 1979 to 0.4 per cent by 2000: J Hawksworth, 'Dude, Where's my Oil Money?' PWC 2008; quoted in A Chakrabortty, 'Dude, Where's My North Sea Oil Money?', *The Guardian*, 13 January 2014.

[192] Public sector current spending remained stable at around 40 per cent between 1979 and 2000: Hawksworth, n191, above.

[193] Hawksworth, n191, above; see also the contribution of I Rutledge to an Institute of Contemporary British History Seminar published as G Staerck (ed), *The Development of North Sea Oil and Gas*, ICBH 2002 ('Staerck, "North Sea Oil and Gas"'), p102.

detriment of future generations;[194] although it should be noted that the failure to create a sovereign wealth fund does not inevitably create inter-generational injustice, as those in receipt of the tax cut may invest the saving in such a way as to allow them or future generations to consume the returns in lieu of the now-lost tax revenues.[195] There is, however, little evidence that this has actually happened in the UK.[196]

The only part of Great Britain which could be said to have accumulated anything akin to a social fund is the Shetland Isles.[197] The local authority – whose chief executive officer has in this regard been said to have been 'far ahead of both the Scottish Office and the Department of Energy',[198] was able to persuade the oil industry to agree to a throughput levy on oil passing through the Sullom Voe terminal. This levy – which the industry agreed to 'as a way of compensating the people for the inconvenience of having the terminal based in Shetland'[199] – has over the years built up into a significant fund, now administered by the Shetland Charitable Trust. The trust supports a broad range of community and arts projects. The fund has net assets of almost £250 million[200] and issued around £10 million in grants during financial year 2014–15.[201]

---

[194] Independent Expert Group to the Calman Commission on Scottish Devolution, *Natural Resource Taxation and Scottish Devolution*, (2009), p13.

[195] A Cummine, above n187 at 33.

[196] S Wren-Lewis, Mainly Macro Blog: *Safe Assets and Sovereign Wealth Funds: Norway, the UK and Oil* (1 February 2013), available at https://mainlymacro.blogspot.co.uk/2013/02/safe-assets-and-sovereign-wealth-funds.html, last accessed 26 May 2017.

[197] Cities such as Aberdeen and areas such as the Cromarty Firth have benefited significantly from the employment and wealth generated by virtue of the oil industry's decision to locate operations there. However, these benefits have been indirect.

[198] See the contribution of I Noble reported in Staerck, *North Sea Oil and Gas*, p79.

[199] Shetland Charitable Trust website, available at www.shetlandcharitabletrust.co.uk/who-we-are, last accessed 19 April 2019.

[200] Shetland Charitable Trust, Statement of Accounts for 2014–15, www.shetlandcharitabletrust.co.uk/assets/files/accounts/Annual-Accounts-for-the-Year-Ending-31-March-2015.pdf, last accessed 19 April 2019, p15.

[201] Ibid, p9.

6.6 BRITISH GOVERNMENT TAKE FROM HYDROCARBON OPERATIONS

Occasional experimentation apart,[202] cash premium bidding – although successfully utilized in other jurisdictions[203] – has not been a feature of the British system. Licences are instead allocated on a discretionary basis, with the extent of the proposed work programme being the principal determinant in the event of competition for the same acreage.[204] At least in theory, this has the benefit of meaning that oil companies do not have to tie up capital in the purchase of acreage, meaning that more capital is available to allow rapid and full development of the acreage let – something that should be in the State's interest;[205] however, Dam, a staunch advocate of cash premium bidding, contends that this argument is overstated and not borne out by empirical evidence.[206] It could be argued that the British discretionary approach has the disadvantage that obtaining acreage on the UKCS is cheap, at least by comparison to systems which *do* utilize premium cash bidding; and developing cheap acreage may seem less of a priority than developing acreage in which significant capital investment has already been made.[207] Arguably, the UK's experiments with cash premium bidding may have occurred at the wrong time in its development as an oil province, when not enough was known about the prospectivity of the areas under offer and the bids received were therefore low. Kemp – while broadly supportive of the UK's discretionary system and noting the arguments against the wholesale use of cash premium bidding upon the UKCS – has argued that the system might have realized 'considerable sums' if it had been selectively used in the prime areas of the North Sea.[208]

---

[202] Some promising acreage in the fourth, eighth and ninth rounds was offered in this way.

[203] Lease sales in the Outer Continental Shelf of the USA, for instance, are undertaken by a process of competitive sealed bidding: A Marino and C Jacob Gower, 'Oil and Gas Mineral Leasing and Development of the Outer Continental Shelf of the United States', IV (2015) *LSU Journal of Energy Law and Resources*, 1 at pp15–16.

[204] G Gordon, *Petroleum Licensing*, in Gordon, Paterson and Usenmez, n18, 4.14 and 4.34.

[205] See, for example, A Beckett, Under-Secretary of State, quoted by A Kemp in *Official History: Vol 1* p248.

[206] Dam, n49 above, p37. Dam argues that the discretionary system also had the effect of decreasing the percentage of foreign participation.

[207] This argument seems fallacious. If someone gives me a car, why should I think it less valuable than an equivalent vehicle I paid for? My focus should be on the value of the asset, not the cost of its purchase. But of course, we do not always think wholly logically, even about major investment decisions.

[208] Kemp, n177, above, at p605.

From the entry into force of the Oil Taxation Act 1975 – which, for all practical purposes, means from the start of production upon the UKCS – state take comprised primarily of a mix of royalty payments,[209] profits on the sale of oil received as a result of direct participation and taxation on the profits of production.[210] As we have seen, direct participation fell out of the equation as a result of the Thatcher reforms of the early 1980s. Royalty, too, was progressively phased out before being abolished altogether.[211] Although having the benefit of being extremely simple to administer, royalty is criticized by economic theorists for its potential regressivity and the distorting effect that it has upon investment decisions;[212] to that extent, its abolition on the UKCS would seem to be wise. However, a system such as the British one, where virtually all of the state take derives from one or another form of tax on profits, leaves the State acutely vulnerable to low oil price.[213] Low oil price means limited or no profits on which to levy tax. In the UK, where as recently as 2011–12 the Treasury received tax revenues of almost £11 billion from oil and gas, the Treasury's revenue flow turned negative in 2014–15, when Petroleum Revenue Tax (PRT) allowances exceeded revenues received by £24 million.[214] As decommissioning gathers speed,[215] and with no clear sign that oil price will increase substantially,[216] the net cash balance may remain neutral or negative for some time.

---

[209] Royalty rate was set at 12.5 per cent *ad valoreum* on the basis that this seemed to be broadly comparable to practice in other petroleum states: A Kemp, *Official History: Vol 1*, p43. Initially, royalty was a significant component of the UK's take: between the late 1970s and the mid-1980s it frequently contributed one-quarter or more of the overall state take: HM Revenue and Customs, *Government Revenues from UK Oil and Gas Production 1968/9–2015/6*, op cit n180.

[210] Acreage fees are also levied, but these make a very small contribution to state take.

[211] Fields receiving development consent after 1 April 1982 were exempted from the requirement to pay royalty. Royalty on all fields was abolished from 1 January 2003. Gordon, 4.29.

[212] See, for example, R Garnaut and A Ross, *Taxation of Mineral Rents* (Clarendon Press 1983) Ch 9.

[213] That vulnerability is rendered more acute by the absence, in the UK, of an oil stabilization fund.

[214] www.gov.uk/government/uploads/system/uploads/attachment_data/file/532649/Table_11_11.pdf, last accessed 19 April 2019.

[215] As we shall see below, significant quantities of decommissioning costs are allowable against tax paid going back many years.

[216] At time of writing, price is currently standing at around $55 per barrel, a significant increase on the $28 per barrel low of January 2016, but well short of the $110 regularly achieved between 2010 and 2014.

Following the discontinuation of state participation and the abolition of royalty, state take now comes only from taxation and area rental fees. Area rental fees make up a small proportion of state take and shall not be further discussed. Tax is by far the most significant element in the UK's state take system. The UK petroleum tax system has, over its relatively short history, been based on sound policy goals: it has sought to ensure that the State was able to obtain a material proportion of the benefits of exploitation of the UK's oil and gas deposits, and to grant incentives for projects that were desirable in policy terms, but were being held back by the fiscal regime. It has, however, been characterized by a relatively high marginal tax rate (albeit ameliorated, to some extent, by the effect of allowances), chronic instability and significant complexity; albeit that, as we shall see below, the reforms recently undertaken in response to the Fiscal Review and (in particular) the further measures taken in response to the price crash of 2014 have gone a long way towards addressing the first and third characteristics.

The specialist tax regime in place for upstream oil and gas operations applies both off- and onshore and is made up of three elements: PRT; Ring-Fence Corporation Tax (RFCT); and Supplementary Charge (SC). Each shall be briefly discussed in turn.

6.6.1 Petroleum Revenue Tax

PRT is a field-based tax payable only on fields developed prior to 1993. PRT was formerly a significant revenue-raiser for the Treasury[217] and was levied at a rate of 50 per cent as recently as the end of 2014, before being cut to 35 per cent as part of the governmental response to the price crash.[218] However, it no longer functions as a means of revenue-raising for the State. The rate of PRT has now been 'permanently set' at 0 per cent.[219] It should be noted, however, that the tax has not been abolished; although it cannot now raise revenue for

---

[217] In every year in the decade between 1979–80 and 1989–90, it contributed in excess of £1 billion to the Treasury, frequently contributing more than a half of the state take from petroleum: www.gov.uk/government/uploads/system/uploads/attachment _data/file/532649/Table_11_11.pdf, last accessed 19 April 2019.

[218] HM Treasury, Budget 2015, available at www.gov.uk/government/uploads/ system/uploads/attachment_data/file/416331/47881_Budget_2015_PRINT.pdf, last accessed 19 April 2019, at para 1.129; Finance Act 2015 s 52.

[219] HM Revenue and Customs, *Policy Paper: Oil and gas taxation: reduction in Petroleum Revenue Tax and supplementary charge*, 2016, available at www.gov .uk/government/publications/oil-and-gas-taxation-reduction-in-petroleum-revenue-tax -and-supplementary-charge/oil-and-gas-taxation-reductionin-petroleum-revenue-tax -and-supplementary-charge, last accessed 19 April 2019; Finance (No 2) Bill 2016 Clause 128.

the State, it has been preserved in order to permit companies to claim losses (including decommissioning costs) against previous payments of PRT. Such losses can be claimed back without limit of time. Thus, PRT will, in the future, exist not as a means of gaining revenue for the State, but as a mechanism by which the State will contribute towards the cost of decommissioning the (typically very large) offshore installations put in place in the 1970s and 1980s.

6.6.2 Ring-Fence Corporation Tax

RFCT has been a feature of the UK tax system from the earliest days of production, having been introduced by the Oil Taxation Act 1975.[220] In the 1970s and 1980s, PRT generally produced significantly more revenue for the state than did RFCT, but from the late 1990s onwards, RFCT has emerged as the more significant of the two taxes.[221]

As the name would imply, RFCT is a particular form of corporation tax which is applied within a 'ring fence' that treats exploration and production activities as a trade distinct from other activities and thus subject to separate taxation. RFCT is calculated in the same way as standard corporation tax. Initially, the rate was the same as that for corporation tax in general, but the rates diverged in 2008. At time of writing, RFCT is levied at 30 per cent, while corporation tax is generally levied at 20 per cent. The effect of the ring fence is to 'prevent taxable profits from oil and gas extraction in the UK and UKCS being reduced by losses from other activities'.[222] It therefore significantly reduces the potential for the manipulation of the oil companies' tax position to the detriment of the State – a problem identified by the Public Accounts Committee in 1972–73.[223] The tax encourages capital investment by means of a 100 per cent first-year allowance for virtually all capital expenditure.[224] Decommissioning costs are also allowable against RFCT, while the decommissioning party continues to be subject to RFCT or within three years of the cessation of ring-fence trade.[225]

---

[220] Oil Taxation Act 1975 s 13.

[221] HM Revenue and Customs, *Government Revenues from UK Oil and Gas Production 1968/9–2015/6*, op cit n180. For instance, in 2001–02, PRT contributed £1.3 billion, while RFCT contributed £3.5 billion. This is broadly typical for this period.

[222] OGA, *Guidance – Oil and gas: taxation*, 2012 updated 2016, available at www .gov.uk/guidance/oil-and-gas-taxation, last accessed 19 April 2019.

[223] See n37 and accompanying text, above.

[224] HM Revenue and Customs, *Oil Taxation Manual, Corporation Tax Ring Fence: First Year Allowances for a ring-fence trade, OT21240*, available at www.gov.uk/hmrc -internal-manuals.oil-taxation-manual.ot21240, last accessed 19 April 2019.

[225] Capital Allowances Act 2001 Ch 13, 'Provisions relative to ring-fence trade'.

It has recently become clear that RFCT (and also SC, as discussed below) has exercised negative effects upon the market for the sale of oil and gas assets, distorting and in some cases frustrating commercial deals and making that market less liquid than would be ideal.[226] The problem is particularly acute in the context of asset sales, as opposed to corporate sales,[227] and is compounded by the fact that it is not always practicable for the parties to structure their transaction as a corporate sale. These difficulties have been caused by the fact that while tax relief is available against the assets' decommissioning costs, it can only be claimed by parties which have accrued a significant tax history within the upstream ring-fence upon the UKCS. While companies that have held assets upon the UKCS for an extended period of time will have built up such a history of tax payments, new entrants to the province will not. As a result, those parties face the prospect of being able to obtain less decommissioning relief than parties with a longer tax history – a relative disadvantage and a disincentive for new players to enter the market.[228] Following representations from the industry, and having regard to the policy of maximizing economic recovery,[229] the government proposes to address this problem by permitting the seller of an oil field to transfer part of its accrued RFCT and SC history to the purchaser. The scheme will be voluntary in nature and the amount of tax history to be transferred will be determined by the parties, subject to regulatory oversight.[230] Enacting legislation is expected in the 2018–19 Parliamentary session.

6.6.3   Supplementary Charge

SC was introduced in 2002 in order to provide the State with a 'fair return',[231] reflecting what the Labour government perceived to be the excessive windfall profits being made by the upstream oil industry at that time. The introduction of SC was controversial, with many arguing that it amounted to a crude 'cash grab' that broke faith with the consultative approach adopted by other limbs

---

[226] See, for example, HM Treasury, *Tax issues for late-life oil and gas assets: A Discussion Paper*, Chapter 3.

[227] An asset sale is when the individual field in question is sold. A corporate sale is when the corporate vehicle is itself purchased, including all assets and liabilities. For further discussion, see N Wisely, 'Acquisitions and Disposals of Upstream Oil and Gas Assets', in Gordon, Paterson and Usenmez, op cit n18, Chapter 12.

[228] HM Treasury, *An Outline of Transferrable Tax History*, 2017, Chapter 1.

[229] Ibid, para 1.10.

[230] Ibid, Chapter 2.

[231] Hansard HC Deb 3 July 2002 vol 388 cc263–303, http://hansard.millbanksystems.com/commons/2002/jul/03/supplementary-charge-in-respect-of-ring, last accessed 19 April 2019, col 297.

of government and would damage investor confidence.[232] However, SC has generated significant revenue for the Treasury, not least because for much of its short life, oil price has been at historic highs. In its 14 years of existence, SC has generated a cumulative total of £27.5 billion of tax revenue, rivalling and occasionally eclipsing RFCT as the principal means by which the oil industry contributes to the Treasury's coffers.[233]

SC is a peculiar fiscal animal: although not technically a corporation tax, it has been designed so as to closely resemble RFCT.[234] The principal difference between the charging bases of SC and RFCT is that in the case of SC, financing costs cannot be taken into account in calculating the profits on ring-fence activities. The government argued that this was necessary in order to prevent companies manipulating the level of their borrowing as between ring-fence and non-ring-fence activities in order to artificially reduce their charge to tax.[235] There are also differences in the allowances available. The 100 per cent capital allowance available for RFCT is not available for SC; however, a number of specific other allowances are available relative to SC only.[236]

When initially introduced, SC was levied at 10 per cent.[237] As oil price increased, it was subsequently raised to 20 per cent[238] before peaking at 32 per cent,[239] at which point it had the effect of taking the UK's marginal rate of taxation up to 81 per cent for fields paying PRT (62 per cent for those which did not). As a result of the Fiscal Review which took place following the Wood Review, the rate of supplementary charge was reduced to 30 per cent.[240] Then, as it became increasingly clear that the price crash of 2014 was not a tempo-

---

[232] UK Offshore Operators Association, Response to Budget Statement, 2002, available at http://oilandgasuk.co.uk/ukooa-response-to-the-budget-statement/, last accessed 19 April 2019; [Hansard debate] Ibid, col, per Mr Flight at col 265 and 276; per Mr Salmond at col 279 and 281, and per Sir Robert Smith at col 282.

[233] HM Revenue and Customs, *Government Revenues from UK Oil and Gas Production 1968/9-2015/6*, op cit n180.

[234] It was introduced by the Finance Act 2002 s 91, which described it as a sum chargeable on ring-fence profits 'as if it were an amount of corporation tax chargeable on the company'. As noted above, the forthcoming scheme for the transfer of tax history will apply to SC in the same way as to RFCT.

[235] HC Deb 03 July 2002 vol 388 cc263–303 per Ruth Kelly at 296.

[236] For instance, investment allowance and cluster area allowance: www.gov.uk/guidance/oil-gas-and-mining-supplementary-charge, last accessed 19 April 2019.

[237] Finance Act 2002 s 91.

[238] Finance Act 2006 s 152. Oil price at the time was around $60 per barrel.

[239] Finance Act 2011 s 7(1). Oil price at the time was around $85 per barrel.

[240] HM Treasury, Driving Investment, 2014, www.gov.uk/government/uploads/system/uploads/attachment_data/file/382785/PU1721_Driving_investment_-_a_plan_to_reform_the_oil_and_gas_fiscal_regime.pdf, last accessed 19 April 2019, para 4.6.

rary dip, and following further representation from industry, the charge was dropped first to 20 per cent[241] and then finally back to 10 per cent.[242]

6.6.4 Conclusion on State Take

At time of writing, it is unclear whether the British government will ever again benefit from the levels of direct state take seen in the mid-1980s or the period from 2005–12. It may be that the best it can hope for will be that tax revenue keeps pace with the allowances associated with decommissioning liability and that the economy continues to benefit from employment and service sector profits, and the taxes thereon.

6.7 BRITISH INVESTMENT PROTECTIONS AND OTHER ENCOURAGEMENTS TO ATTRACT INVESTORS

The British government has entered into a range of international investment protection arrangements and engaged in a range of policy activities designed to encourage and protect investment. These shall be discussed in turn.

6.7.1 Energy Charter Treaty 1994

The UK has ratified the Energy Charter Treaty;[243] as a result, it is bound by that Treaty's provisions on investment promotion and protection.[244] These – intended, like the other provisions of the Treaty, to 'promote long-term co-operation in the energy field, based on complementarities and mutual

---

[241] Finance Act 2015 s 48.

[242] HM Revenue and Customs, Policy Paper, op cit n219; Finance (No 2) Bill 2016, clause 54.

[243] The Energy Charter Treaty 1994: for the official text and further information, see the Energy Charter Treaty microsite, www.energycharter.org/process/energy-charter -treaty-1994/energy-charter-treaty, last accessed 19 April 2019.

[244] The UK is also a signatory to the International Energy Charter 2015, an instrument which has been signed by a larger and more diverse group of countries than the Energy Charter Treaty: for the official text and further information, see the International Energy Charter microsite, www.energycharter.org/process/international -energy-charter-2015/overview/, last accessed 19 April 2019. As the International Energy Charter's recitals make clear, it contains only political commitments, not legal obligations; consequently, its provisions shall not be further considered here. However, it is worth noting that the Energy Charter Treaty grew out of the non-binding European Energy Charter; it is possible that the International Energy Charter will in time give rise to an instrument with legal effect.

benefits'[245] – are contained in Part III of the Treaty. The word 'investment' is given an expansive definition, encompassing 'every kind of asset, owned or controlled directly or indirectly by an Investor'.[246] Among other things, it specifically includes tangible and intangible property, all forms of equity participation in a company or business enterprise, energy licence or permit rights and intellectual property rights.[247] The Treaty would appear to apply to investments located within the UKCS, as well as investments located onshore or within the territorial sea.[248]

The Treaty provides a range of provisions in relation to investment, some more concrete than others. The obligations relative to the making of investment are contained in Article 10. Article 10(1) provides a high-level obligation that the member state will 'encourage and create stable, equitable, favourable and transparent conditions for Investors of other Contracting Parties to make Investments in its Area'. This specifically includes an obligation to accord investors of other Contracting Parties 'fair and equitable treatment'. Article 10(2), read together with Article 10(3), provides that, relative to the making of investments in their area, Contracting Parties must 'endeavour' to provide investors from a Contracting Party with treatment which is no less favourable than that accorded to any other class of investor. The non-binding nature of this obligation is underlined by the fact that (1) the Charter envisages that the making of investment shall be the subject of a supplementary treaty;[249] (2) each Contracting Party undertakes to endeavour to minimize the number of exceptions it makes to the no less favourable treatment rule,[250] and to 'progressively remove' investment restrictions affecting Investors of other Contracting Parties;[251] and (3) Contracting Parties are empowered to notify to the Secretariat a voluntary commitment to accord no less favourable treatment

---

[245] Art 2.

[246] Art 1(6).

[247] In view of the definition's length, this is a simplified account. See ibid for the full definition.

[248] 'Area' is defined as territory under a Contracting Party's 'sovereignty' (Art 1/10(a)) and so as to include the sea, seabed and its subsoil with regard to which that Contracting Party exercises sovereign rights and jurisdiction (Art 1.10(b)). The UKCS would seem to be encompassed by this definition by virtue of the State's exercise of sovereign rights thereover.

[249] Art 10(4). It was envisaged that this Treaty would be concluded by 1998. It has not yet materialized.

[250] Art 10(5)(a).

[251] Art 10(5)(b). In the UK, the ability to restrict investment in the upstream oil and gas sector is in any event significantly curtailed by the Hydrocarbons Licensing Directive, discussed above.

to investors of other Contracting Parties, which commitments will become binding when listed in the relevant Annex to the Treaty.[252]

The provisions relating to the protection of investments are more concrete than those which relate to the making of investments. Article 13(1) provides that investments of investors of a Contracting Party shall not be expropriated[253] unless such expropriation is in the public interest, non-discriminatory, carried out under due process of law and accompanied by the payment of prompt, adequate and effective compensation. Such compensation must reflect fair market value, be provided in freely convertible currency and include interest at a commercial rate. The investor must be provided with a prompt right of review by a judicial or other competent and independent authority.

The Treaty also contains provisions relative to the employment of key personnel,[254] the free transfer of capital, returns, compensation and payments,[255] and the payment of compensation for loss suffered as a result of war or other armed conflict, state of national emergency, civil disturbance.[256]

6.7.2   Bilateral Investment Treaties

Quite apart from the protections afforded by the Energy Charter Treaty, overseas investors in the UK could potentially benefit from the provisions of a Bilateral Investment Treaty. The UK is party to a large number of these,[257] including some with states which are not party to the ECT, but which are the domicile of investors active on the UKCS – for instance, China[258] and the United Arab Emirates.[259] It is dangerous to generalize about the nature

---

[252] Art 10(6)(b). No such commitments have yet been made.

[253] More fully, 'nationalised, expropriated or subjected to a measure or measures having effect equivalent to nationalisation or expropriation'.

[254] Art 11 provides the obligation to 'examine in good faith' requests to employ within the Contracting Party key personnel from outside its territorial area.

[255] Art 14.

[256] Art 12 provides that in such circumstances, compensation shall be made on a most favourable treatment basis.

[257] A search using the United Nations Conference on Trade and Development online service, available at http://investmentpolicyhub.unctad.org/IIA/CountryBits/221, last accessed 19 April 2019, discloses that the UK is a party to 110 Bilateral Investment Treaties, most of which are in force; and over 70 Treaties with Investment Provisions.

[258] Agreement between the Government of the Kingdom of Great Britain and Northern Ireland and the Government of the People's Republic of China concerning the Promotion and Reciprocal Protection of Investments with Exchanges of Notes, 1986, Cmnd 9821.

[259] Agreement between the Government of the United Kingdom of Great Britain and Northern Ireland and the Government of the United Arab Emirates for the Promotion and Protection of Investments, 1992, Cm 2535,

of the protections provided by these Treaties; the detail[260] of their provisions vary, with older Treaties generally being less sophisticated than more recent ones. However, all examples reviewed by the author have contained provisions broadly similar to those of the Energy Charter Treaty. The European Convention on Human Rights, incorporated into domestic law by the Human Rights Act 1998, could also provide an investor with a remedy in the event of, for example, expropriation.[261]

6.7.3 Other Measures Intended to Encourage Investment

In addition to the above, Britain has sought to encourage investment by trying to establish itself as an attractive place to do business. Through attendance at international trade fairs and the development of the Promote UK website, which acts as a shop window for the UK industry and collates the key elements of the offshore regulatory regime in one place,[262] the government has tried to encourage fresh entrants to invest in the UKCS by painting a picture of the UK as a stable[263] and straightforward place in which to do business, even going so far as to helpfully advertise to investors particular plays that might be of interest.[264] Innovative licensing models have also been developed, tailoring the UK licence to meet particular circumstances in an attempt to balance the legitimate interests of both the licensee and the State.[265] More recently, the Wood Review sought to revitalize the UKCS by reorienting commercial behaviours

---

[260] For instance, although it is clear that each of the examples given above applies to the UKCS, the wording used to achieve this result differs markedly: for the UK-UAE BIT, see op cit n259 Art 1(f)(i); for the UK-China BIT, see the simpler (but it would seem, equally effective) wording at op cit n258 Art 1(2).

[261] Protocol 1 Article 1 to the European Convention on Human Rights provides:
Every natural or legal person is entitled to the peaceful enjoyment of his possessions. No one shall be deprived of his possessions except in the public interest and subject to the conditions provided for by law and by the general principles of international law.

[262] DECC, Promote UK 2015, available at https://itportal.decc.gov.uk/web_files/promote/2015/index.htm, last accessed 18 April 2019.

[263] DECC's attempts in this regard have at times been somewhat undercut by the behaviour of HM Treasury: see nn100–107 and associated text, above.

[264] DECC, Promote UK 2015. See the information available through the buttons marked 'North Sea Opportunities', 'West of Britain Opportunities' and 'Selected Play Summaries'.

[265] The Frontier Licence was designed to be of particular interest to the major player seeking exclusivity over a large area of unproven territory, while the Promote Licence was designed to encourage the small player: see further G Gordon, 'Petroleum Licensing', in Gordon, Paterson and Üşenmez, op cit n18, paras 4.54–4.68.

and regulatory strategy.[266] The intention behind this was ultimately to stimulate further and smarter investment, and thereby to maximize economic recovery; however, it is at least arguable that the uncertainties attending such a radical overhaul of the governance of the UKCS may have had a chilling effect upon investment, at least in the short term.

6.8   CONCLUSION

If the development of the offshore British oil industry is a story, it is one of evolution interspersed by the occasional revolution, the first of which was the Labour Party's control-oriented reforms of the 1975, the second of which was the dismantling of direct state participation by the Thatcher administration and the most recent of which was the introduction of the OGA following the Wood Review. If it is a picture, it needs to be viewed from a number of different perspectives in order to be fully seen. First, it should be viewed in the light of the broader contest of ideologies that took place between the 1960s, 1970s and 1980s, with the political left consistently (if sometimes cautiously) advocating some measure of state control and participation while the political right, after some initial hesitation, moved decisively in favour of deregulation and private enterprise. Thus, although at times it seemed that a significant participatory role might emerge for the State, ultimately this withered away in the face of Thatcherite ideology. As a result, the UK – which once looked like it would stand alongside Norway as a further example of a North Sea system of statist oil and gas development – instead came to stand alongside the USA as one of the clearest examples of a jurisdiction where the task of the development of the nation's petroleum has been given over to private industry. Belief in small, non-interventionist government, faith in the market's ability to self-regulate and a general tendency to regard oil and gas not as a thing apart, but merely as another industrial sector led to the current lack of state participation and may also have contributed to the lack of a highly developed resource management strategy such as was formulated in Norway. It may also go some way towards explaining why the UK lacks other special features commonly seen in other provinces, such as a social fund.

Regard must also be had to the difficult economic situation in which Britain found itself in the pioneering days of the industry, the rapid speed at which the picture unfolded and the resultant lack of a phase of detailed strategic planning prior to the advent of exploration and production activities. History has at times cast a long shadow, with key features of the legal and fiscal system

---

[266]   See the discussion of the Wood Review and its implementation at section 6.2.2.3, above.

having to be designed in a hurry and then requiring to be fixed – sometimes on numerous occasions – as a result of less-than-optimal initial design. These difficulties were compounded by the fact that some design decisions – most notably, those relating to the legal nature of the licence – served to hinder subsequent change. The UK's production licence, with its peculiar hybrid nature, is not a model to be emulated. It is important to note, however, that the licence does not constitute the entire relationship between the State and the oil company. Change has been regularly effected to, for example, the tax regime and regulatory law, with the implementation of the Wood Review providing an example of how the difficulties caused by a sub-optimal licensing system can be addressed through the creation of new statutory obligations.

Economic factors again intruded at a much later stage in the development of the province, with Treasury-led tax grabs hindering DECC's policy objectives, while funding cuts in the 2000s put significant pressure on DECC's budget at the precise time when the province was becoming more diverse and difficult to govern. As a result of the cumulative effect of political ideology, the difficulty of the task and a lack of resources, a *laissez-faire* or light-touch approach to regulation developed whereby the State adopted an essentially passive role, acting less as a proactive designer of a holistic development policy and more as a reviewer of individual development plans. The Wood Review's introduction of the obligation to maximize economic recovery of petroleum across the UKCS as a whole and its putting the UK's resource management system in the hands of a more interventionist, specialist regulator marks a dramatic shift in policy at a very advanced stage of the UKCS's development. While it is too early to say with certainty how effective the new approach will be, the experience of the OSO provides a historical example of an excellent idea which developed too late to be fully effective. It is to be hoped that the value that Wood rightly noted could be unlocked through a change in regulatory approach and commercial behaviours can be realized as the UKCS becomes progressively more mature.

From the standpoint of the orderly management of the province, many of the lessons to be learned from the study of the UK might be thought to be negative ones. As we have seen, significant difficulties have been created by the dash towards early offshore production, the absence of a clear and consistent long-term strategy by the State's unwillingness at times to treat oil and gas as a special sector of the economy. However, there are positive dimensions, too. If the UK can be criticized for not systematically planning out its oil and gas activities in advance and only learning from experience, at least it *has* learned from experience. Thus the UK's track record for finding ingenious solutions to practical problems is strong. Cooperation between the industry and the government in areas of mutual interest; intelligent tax reforms designed to solve particular problems, remove perverse incentives and increase liquidity in the asset-transfer market; and the (albeit belated) recognition of the need for a coherent strategy to maximize economic recovery: these stand as examples of the UK's positive legacy.